Design in Object Technology
"Class of 1994"

Series on
Object-Oriented
Design

Alistair Cockburn

Design in Object Technology: Class of 1994

© Alistair Cockburn 2021, all rights reserved

ISBN 978-1-7375197-0-6
Humans and Technology Press
32 W 200 S #504
Salt Lake City, UT 84101

Design in Object Technology: Class of 1994

Preface to the 2021 Reprinting

In 1991 I was hired by the IBM Consulting Group to create a methodology for their object-technology projects. An early "agile" methodology, it's emphasis was on incremental development, requirements in use cases, and design using responsibilities.

We applied the methodology in 1994 on a fixed-price, fixed-scope project that integrated COBOL programs with a sizable Smalltalk application via a relational database. Bid as a $10M, 18-month, 50-person project, it delivered on time at a cost of about $15M. The client was happy with the result and the system was still i being maintained ten years later, so it was considered a successful project. The project is written up in detail as "project Winifred" in my 1997 book *Surviving Object-Oriented Projects*.

At the start of the project, I gave a week-long course to the entire team. it covered incremental development, use cases, responsibilities, an early hexagonal architecture, methodologies, whatever they would need to function on the project.

This book is no more and no less than slides from that course, all 214 of them. To honor its historical purpose, I have made no changes to the slides. What you see is what I taught back then.

This book is possibly of greatest interest to those people who were practicing object-oriented design back in the 1990s. They will be interested to see how I presented topics that were current back then. Newer designers might find it interesting to see how we talked about things back then, some might find new ideas new to them that may help them in their designs.

I hope you enjoy this historical educational artifact.

Alistair Cockburn

Design in Object Technology: Class of 1994

Design in Object Technology: Class of 1994

Design in Object Technology

Series on Object-Oriented Design

Rehydrated from the 1994 week-long course in 2021!
--Alistair Cockburn

Humans and Technology
Alistair Cockburn   © 1994-2021 Alistair Cockburn
section (slide)
L1-00 (1)

Design in Object Technology (Course Outline)

e0: Bank Design Map OO Structure		Responsibilities Scenarios		Recursive design Incremental Devt.	(e5: Pricing)
Objects msgs	e1: dates	e4: Human Coffee Machine		e5: Pricing	Reuse Methodologies
Classes, Inheritance	e2: Pizza			Documentation Frameworks	Survival Tips
	e3: Bank	Use cases	(e4: Coffee machine controller)		e6: Account Log
Interaction diagrams		Models & shields		(e5: Pricing)	Stories Questions

Humans and Technology
Alistair Cockburn   © 1994-2021 Alistair Cockburn
section (slide)
L1-00 (2)

Design in Object Technology: Class of 1994

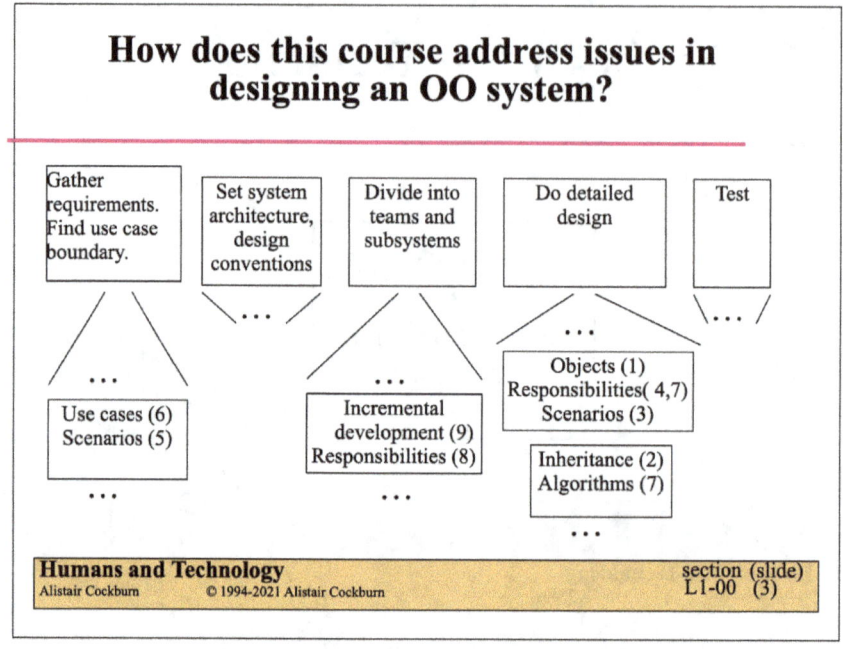

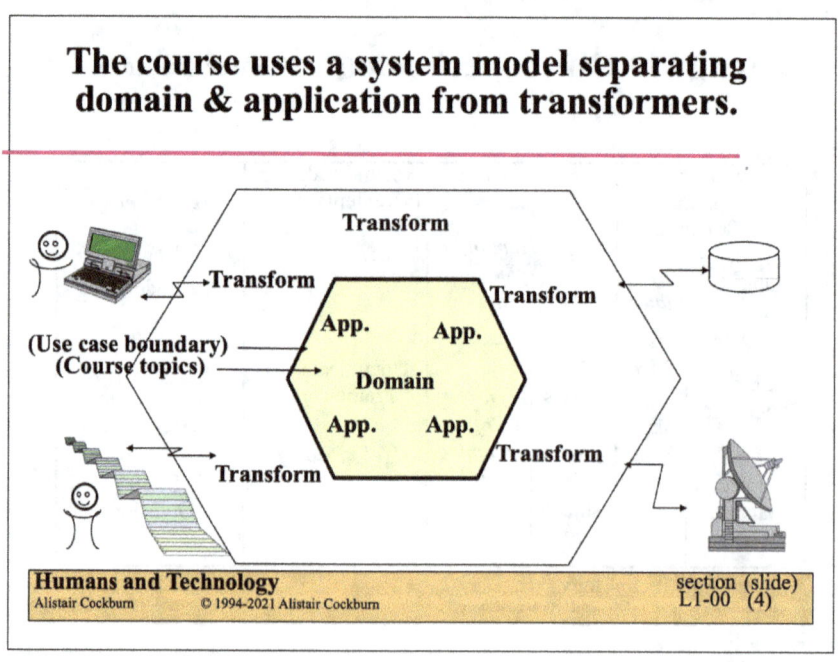

Design in Object Technology: Class of 1994

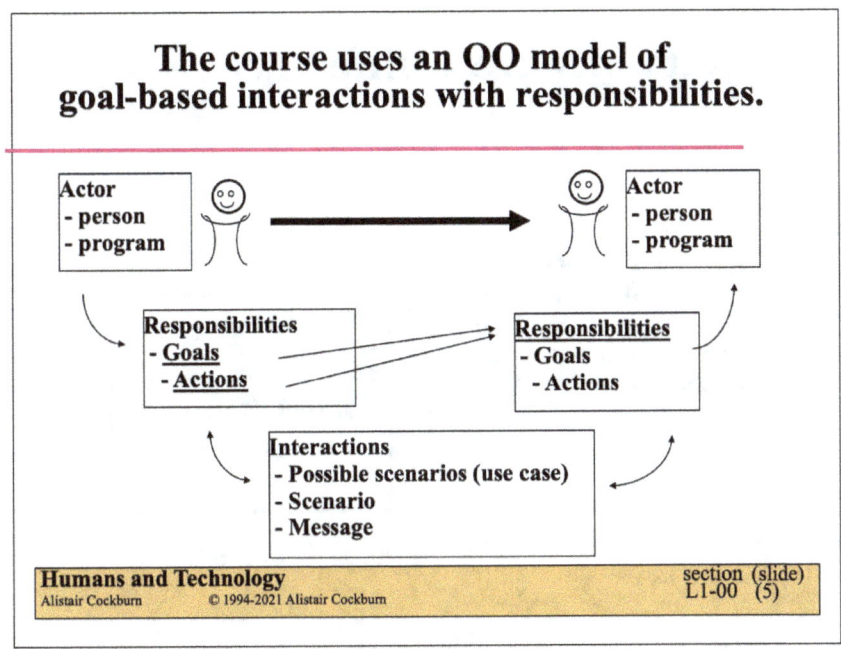

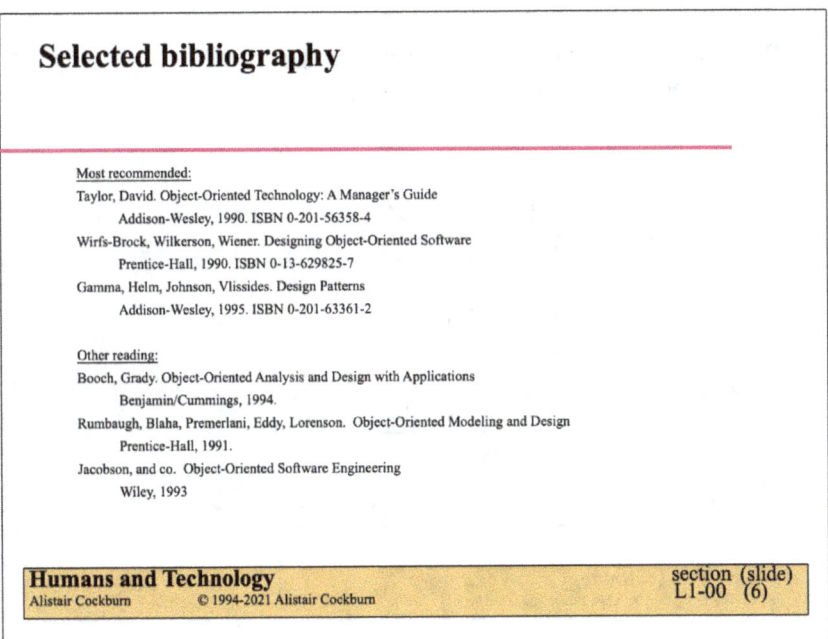

Design in Object Technology: Class of 1994

Object Software's Difference

How does an OO application differ from what a good, modern programmer would write?

Series on Object-Oriented Design

Humans and Technology
Alistair Cockburn   © 1994-2021 Alistair Cockburn
section (slide) L1-01 (7)

The essence of an application is a matrix of functions and data structures.

	□	○	/	AB	✼✼
create	☐	☐	☐	☐	☐
alter	☐	☐			☐
shift	☐	☐	☐	☐	☐
volume?	☐	☐	☐		
stats?	☐	☐	☐	☐	☐

Humans and Technology
Alistair Cockburn   © 1994-2021 Alistair Cockburn
section (slide) L1-01 (8)

Design in Object Technology: Class of 1994

In modular programs, the matrix elements are kept separate, connected by switches at the functions.

	□	○	/	AB	✻✻
create					
alter					
shift					
volume?					
stats?					

Humans and Technology
Alistair Cockburn   © 1994-2021 Alistair Cockburn
section (slide) L1-01 (9)

The switches connect separated modules, limiting change and damage zones.

```
volume ( composite )
  for i = 1 to composite.size do
    if composite.i = □  then volume = volume + volumeOf □
    if composite.i = ○  then volume = volume + volumeOf ○
    if composite.i = AB then volume = volume + volumeOfAB
    if composite.i = ✻✻ then volume = volume + volumeOf✻✻
```

volumeOf □ ...

volumeOf ○ ...

volumeOf ✻✻ ...

Humans and Technology
Alistair Cockburn   © 1994-2021 Alistair Cockburn
section (slide) L1-01 (10)

Design in Object Technology: Class of 1994

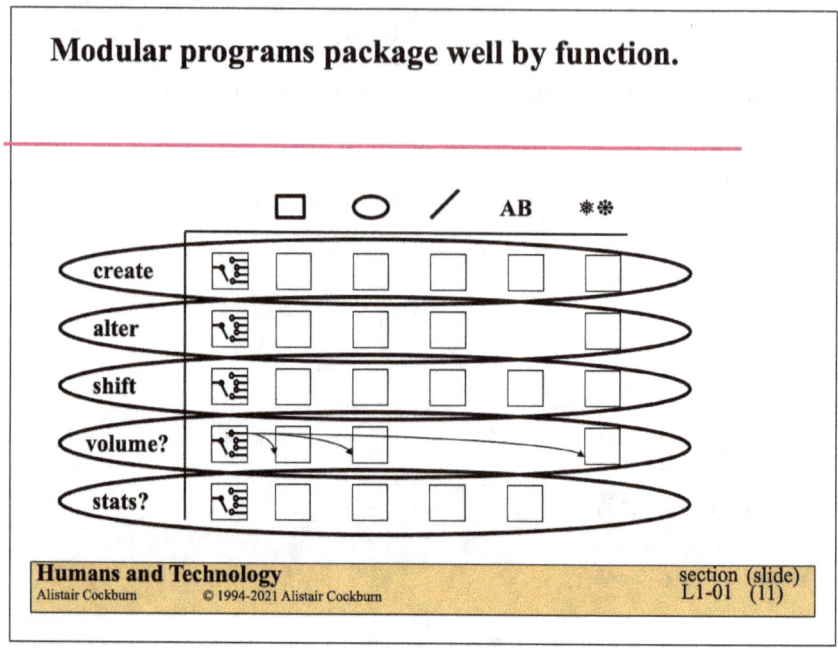

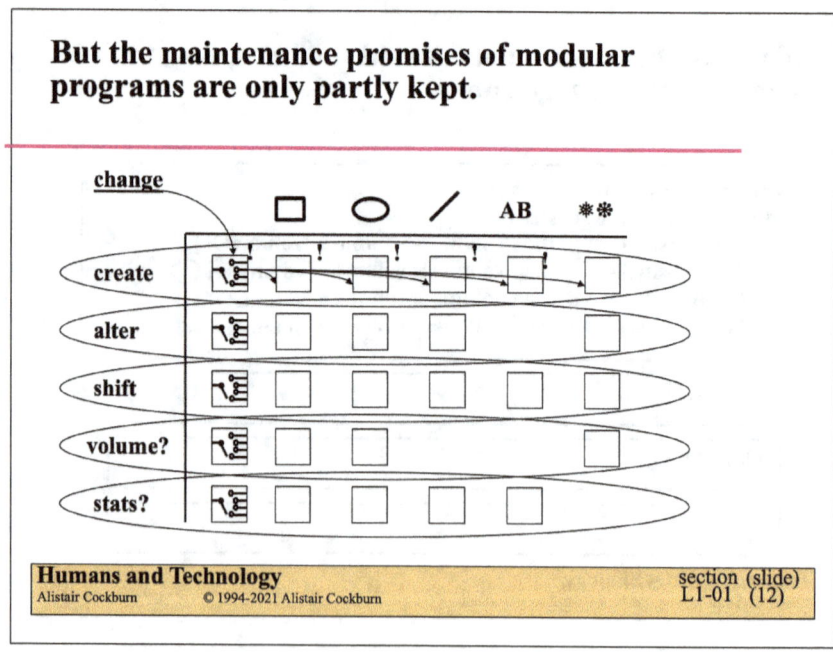

Design in Object Technology: Class of 1994

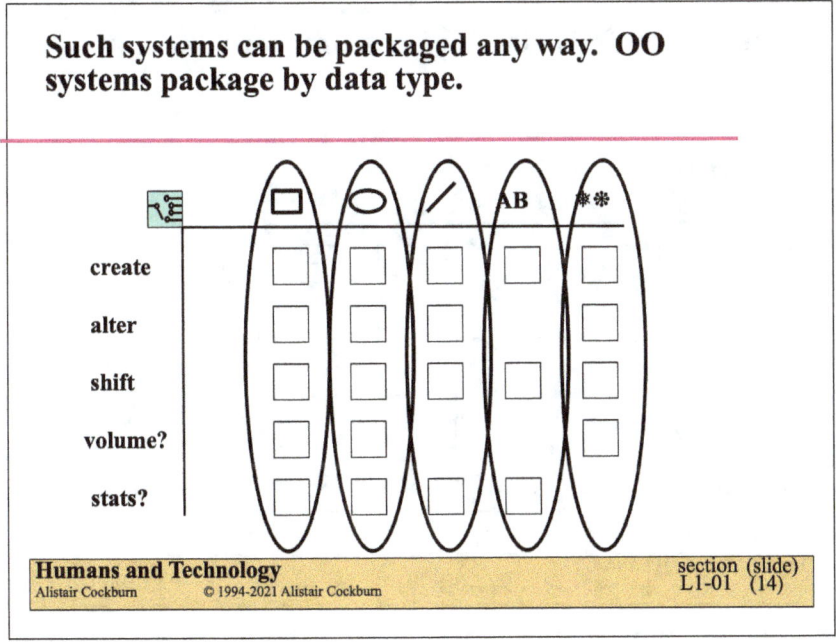

Design in Object Technology: Class of 1994

Isolating the common switch reduces and simplifies the program...

```
composite.volume ( )
 foreach item in composite do
   volume = volume + item.volume( )
```

☐volume( ) ...

○volume( ) ...

❋❋.volume( ) ...

Humans and Technology
Alistair Cockburn   © 1994-2021 Alistair Cockburn
section (slide) L1-01 (15)

...but makes following the program more difficult.

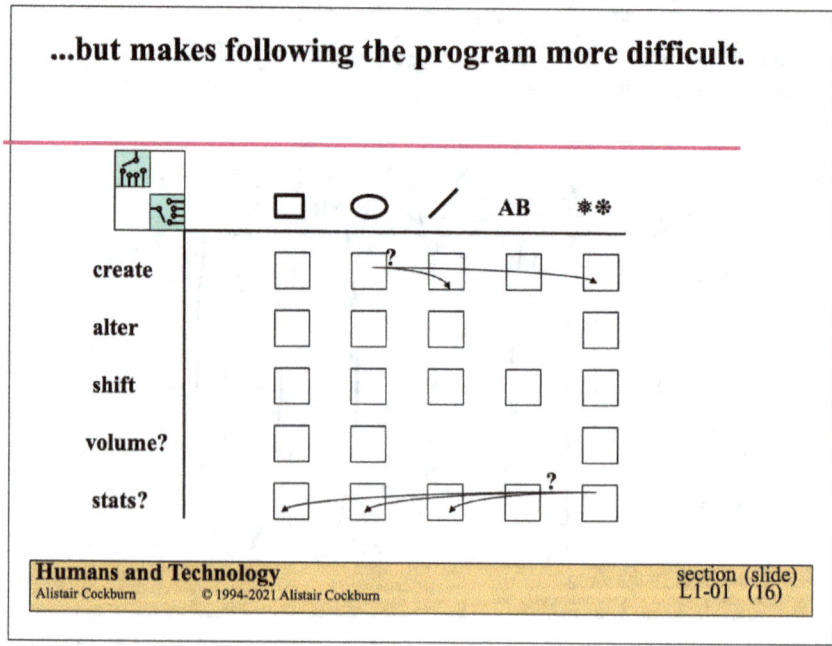

Humans and Technology
Alistair Cockburn   © 1994-2021 Alistair Cockburn
section (slide) L1-01 (16)

Design in Object Technology: Class of 1994

"OO": switching mechanism; less code, unconnected; modularized by object.

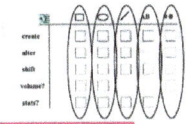

1. OO programs require external mechanisms to carry the common switching code.
   - ...in C++ it is provided by the compiler.
   - ...in Smalltalk it is provided by the run-time.
2. OO programs facilitate modularization by data type (the "objects").
3. OO programs should be shorter, but control flow harder to understand.

Humans and Technology
Alistair Cockburn   © 1994-2021 Alistair Cockburn
section (slide) L1-01 (17)

End of "the cause of the differences"...

Epilogue:
  Localizing change zones is still a key design goal.
  Selecting suitable modules is the design topic.

Questions for Consequences
  How does the different structure affect application development?
  Where can OO (not) be used?

Humans and Technology
Alistair Cockburn   © 1994-2021 Alistair Cockburn
section (slide) L1-01 (18)

Design in Object Technology: Class of 1994

Design with Objects and Messages

Series on Object-Oriented Design

Definition 1: An object is a "thing", with all the state and services that come with it.

What is a telephone?
... a "thing" that cost money, has a number, and which we use to make and receive calls.

In your program, what will "a Telephone" be?
... an "object" with selected properties and services.
- may have a cost (if you are a retailer)
- may dial for you
- likely to "know" its phone number

Design in Object Technology: Class of 1994

Metaphor 1: Think of an object as a "thing" with a personal secretary to handle your commands.

"<u>Dial this number</u>", you command...
    (the secretary takes the phone off hook, dials...)
    the object takes the phone off hook, dials...
"<u>How much did you cost</u>?", you ask...
    (the secretary looks into the phone's records...)
    the object looks into its records...

An object accepts both queries and commands, sometimes with return values, sometimes not.

<u>You</u>	<u>It</u>
How much did you cost?	"$50"
What is your number?	"943-8484"
How long did that last call take?	"16 min."
Redial.	(no comment)
Transfer control to my handset.	(no comment)
Add 3-party to your capabilities.	(no comment)

Design in Object Technology: Class of 1994

Sometimes an object does its work alone, sometimes it relies on another object.

How much did you cost?   "$50"

Redial
--->   hm, what was that last number?   943-8484
--->   to dialer: "9"..."4"..."3"..."8"..."4"..."8"..."4"

How long did that last call take?
--->   hm, what was that last call?   LastCall
--->   to LastCall: How long were you?   "16 min"
--->   "16 min"

"Object design" is the design of how much knowledge and control goes into each object.

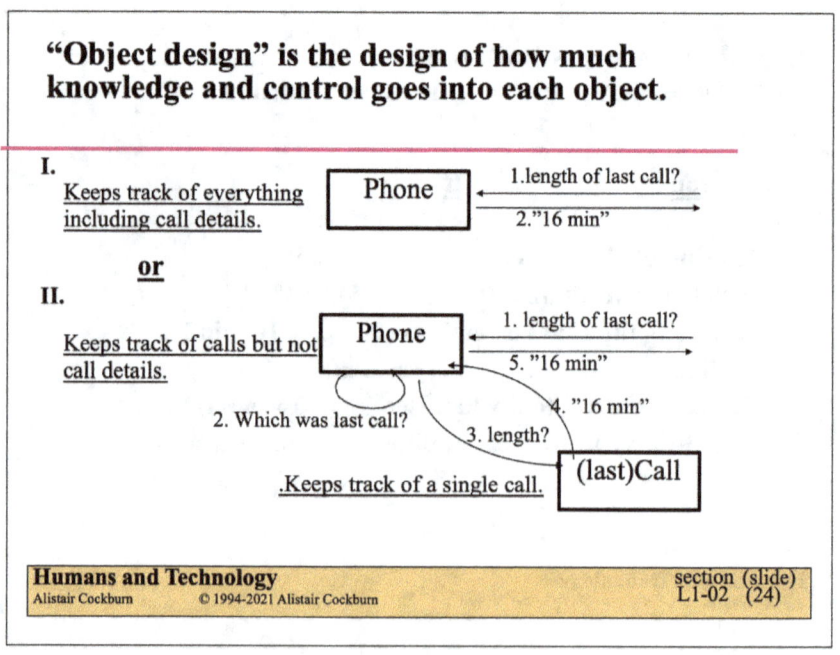

I.
Keeps track of everything including call details.

      or

II.
Keeps track of calls but not call details.

Keeps track of a single call.

Design in Object Technology: Class of 1994

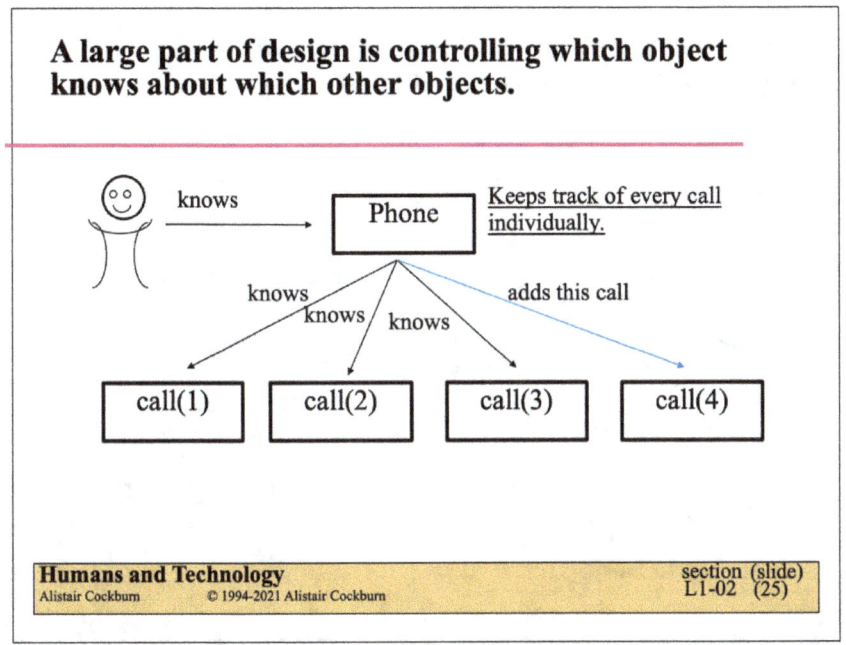

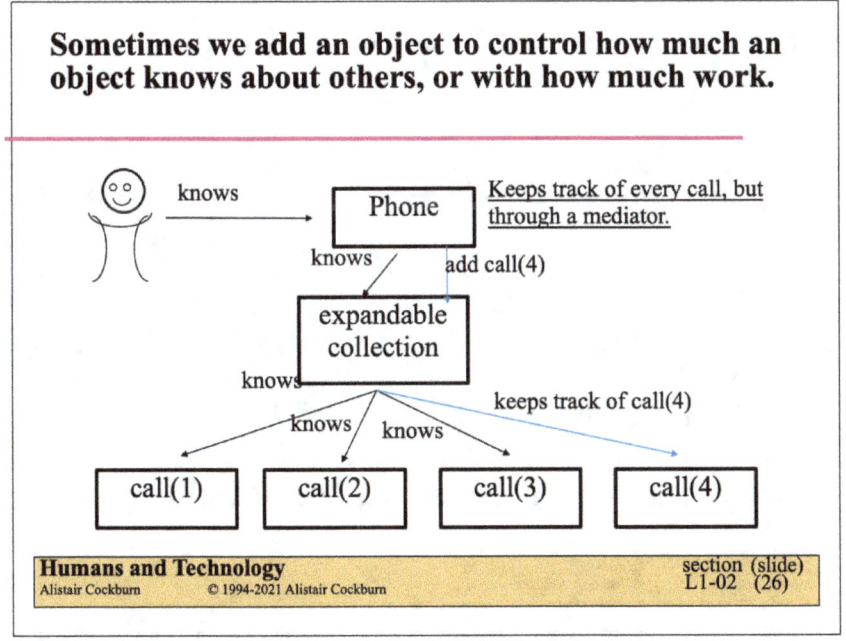

Design in Object Technology: Class of 1994

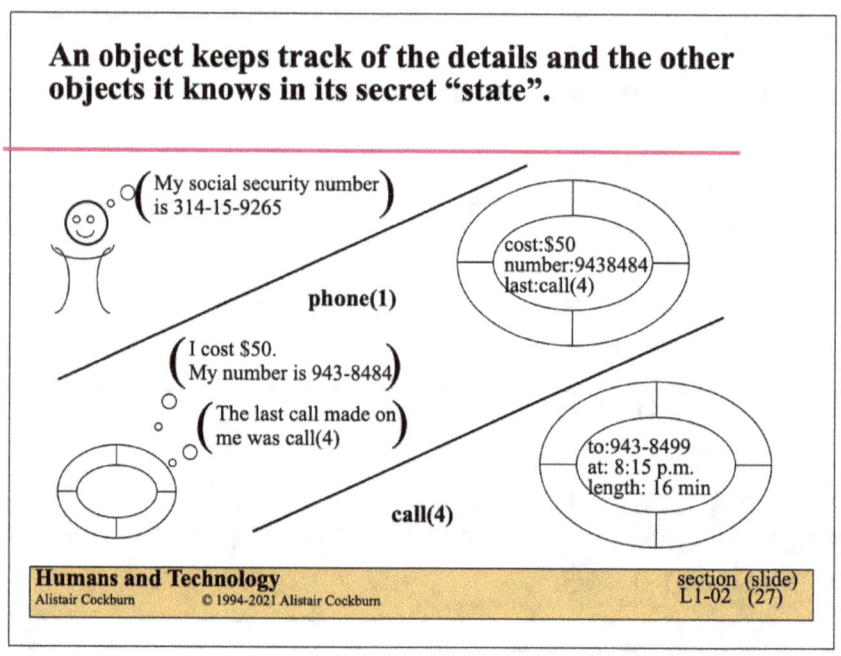

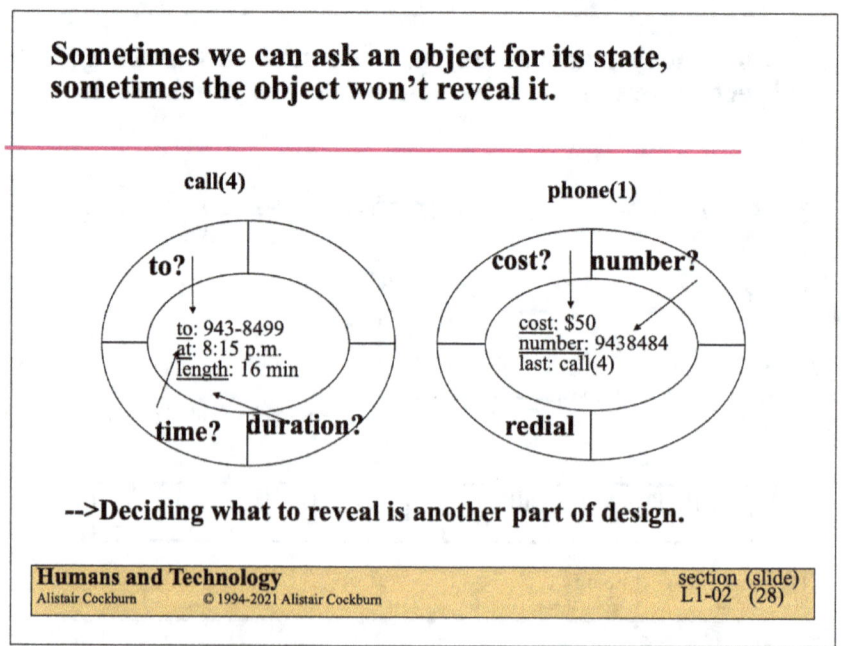

Design in Object Technology: Class of 1994

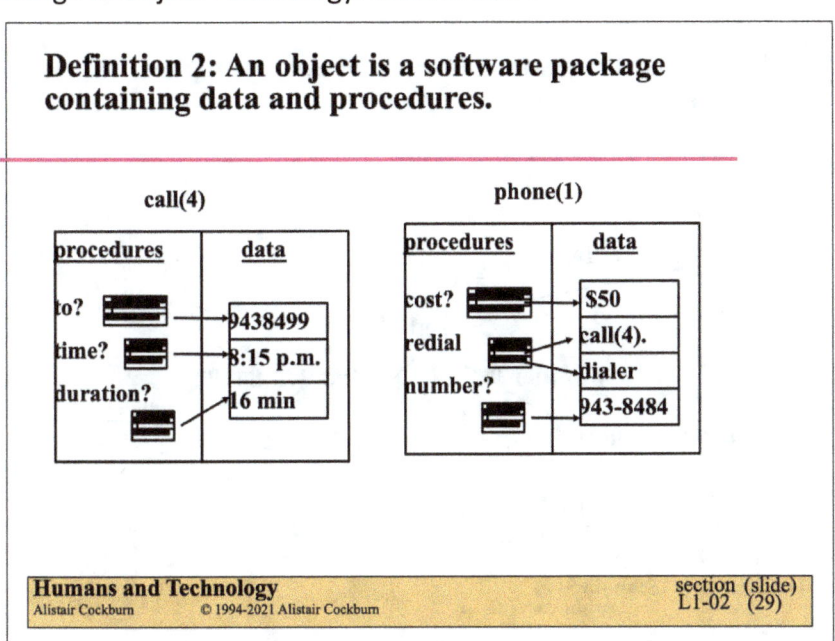

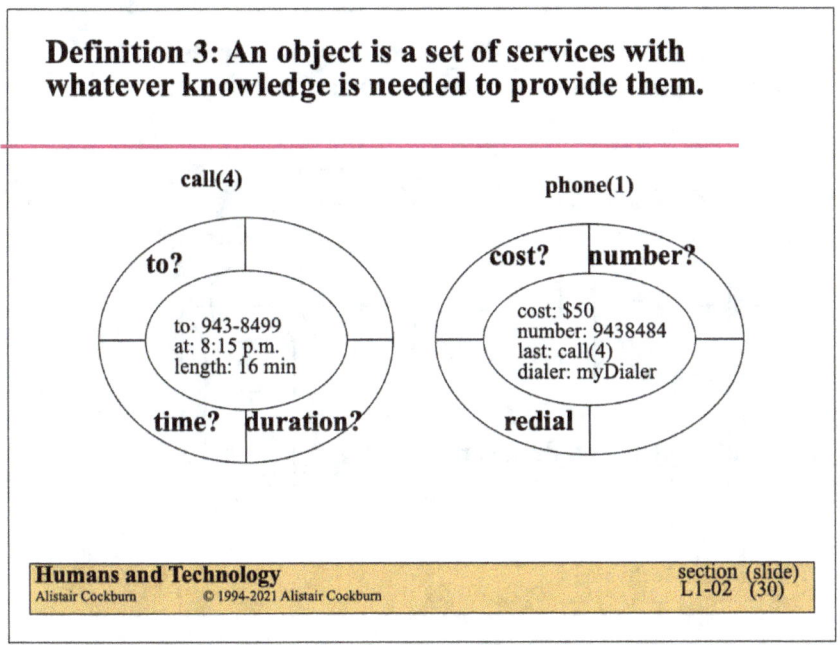

Design in Object Technology: Class of 1994

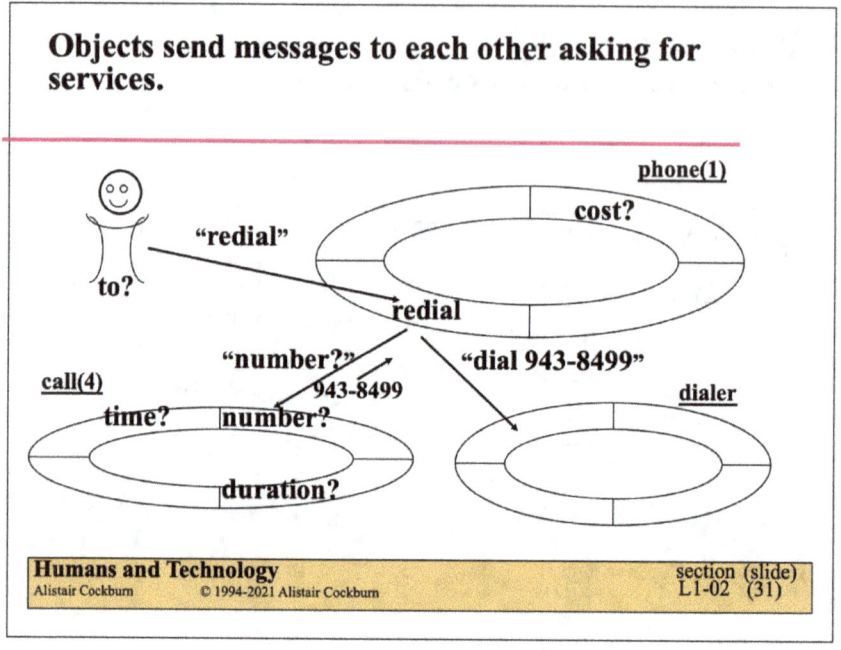

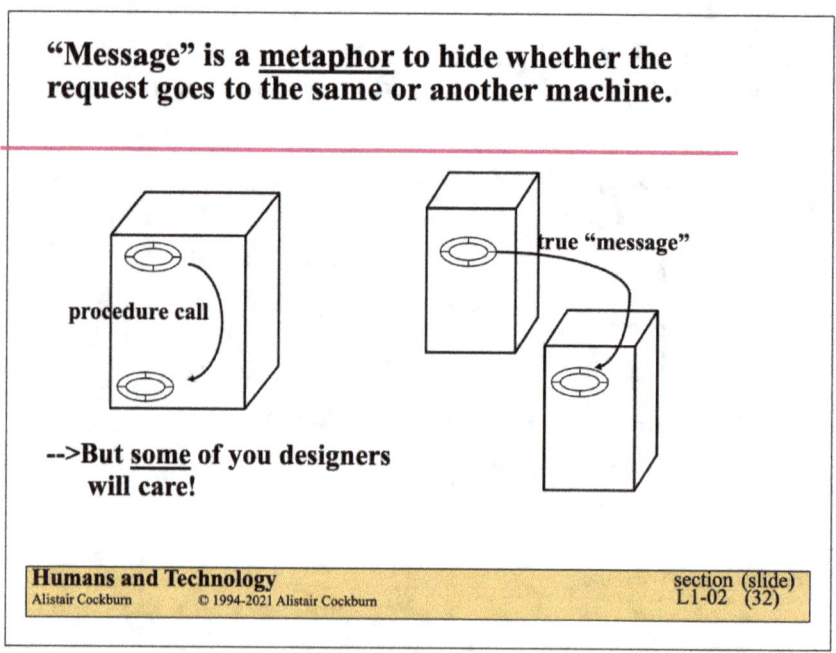

Design in Object Technology: Class of 1994

Objects reviewed: service, knowledge, message

Service	Knowledge	Message
method	instance data	invocation
behavior	instance variables	delegation
function	member data	(call)
procedure	state	
	(attribute)	

<u>Polymorphism</u>: a long way to say that different objects can be handed the same query or command, and handle it as appropriate.

A benefit of Object is it provides the designer the spectrum from all data to all behavior.

The average object has some data, some procedure.

A "Value" is data that must be associated. together.
- similar to a "data structure"
- e.g. a date, a check.

A "Script" is mostly behavior.
- like a procedure or rule base

Objects allows the designer to choose the amount of data and the amount of behavior.

Design in Object Technology: Class of 1994

A hazard to beginners is to model "functions" as objects, when they should be services of objects.

A function can be a legitimate object.
  e.g.: multiplication, transferring money.
An object <u>should</u> have State and Services
  - what is the state of the transfer?
  - checkpoint the multiplication!
Experienced developers know how to discover functions.
  e.g.: printing
Beware functions (services) masquerading as objects!
  Hint: does it have more than 1 service?

Humans and Technology
Alistair Cockburn   © 1994-2021 Alistair Cockburn
section (slide) L1-02 (35)

**Designer's review and Rule 1:
"do? (delegate?), know?, reveal?, locate?".**

1. "What should THIS object do?"
2. "Should it do it all or delegate some of it?"
   • which parts get delegated...to whom?
3. "How much it know?"
4. "What should it reveal?"
5. "What pieces should be co-located?"
   (advanced topic)

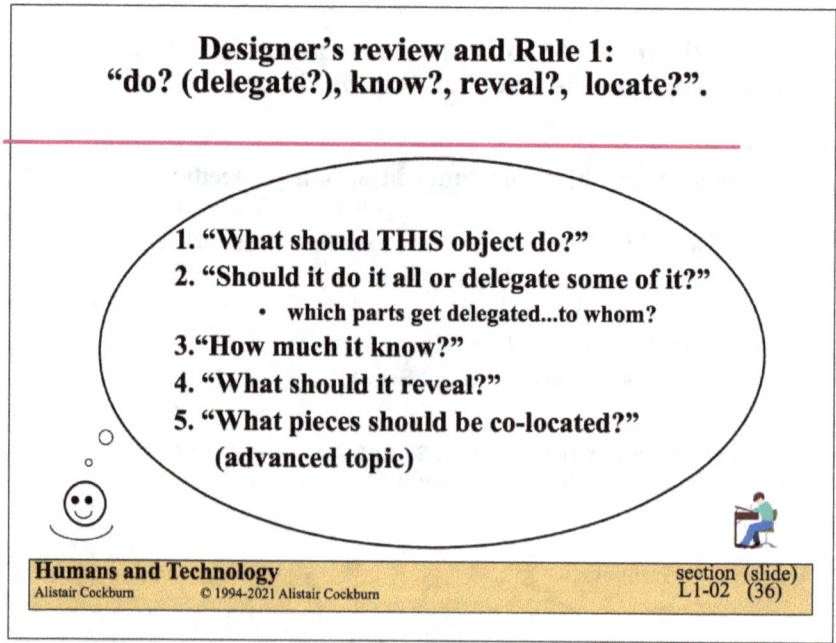

Humans and Technology
Alistair Cockburn   © 1994-2021 Alistair Cockburn
section (slide) L1-02 (36)

Design in Object Technology: Class of 1994

Designing with Classes and Inheritance

Series on Object-Oriented Design

Humans and Technology
Alistair Cockburn   © 1994-2021 Alistair Cockburn
section (slide) L1-03 (37)

Definition 1: A "class" is a repository for the description of common properties of objects.

Non-OO
add, take, open, close → acct. 703, acct. 702, acct. 701

Object-based
- acct. 701: add, take, open, close
- acct. 702: add, take, open, close
- acct. 703: add, take, open, close

Class-based
add, take, open, close ← acct. 703, acct. 702, acct. 701

Humans and Technology
Alistair Cockburn   © 1994-2021 Alistair Cockburn
section (slide) L1-03 (38)

Design in Object Technology: Class of 1994

Definition 2: "Inheritance" is the sharing of commonality across classes.

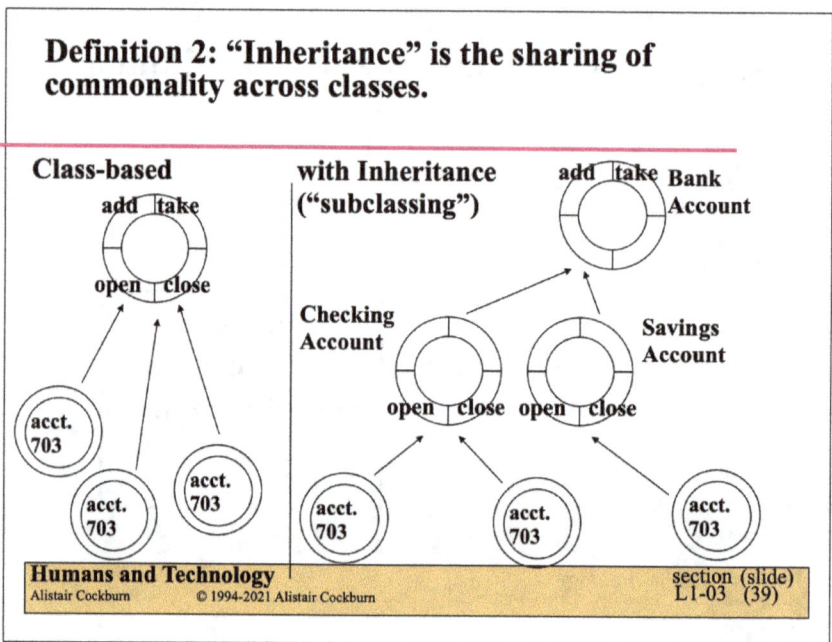

Classes come automatically with some languages, not with others, but can be built with hard work.

Modula supports objects, not classes.
Ada supports objects, not classes.
Smalltalk, Eiffel support classes and objects.
C permits construction of objects and classes.
- for example, C++ or Objective-C or C+@

Languages that have classes usually have inheritance, and vice versa.
- They are typically either Object-Oriented or not.

You can do OO design in any language (with work).

Design in Object Technology: Class of 1994

"Inheritance" in the problem space does not necessarily match inheritance in the design space.

Is a rectangle a subclass or a square, or vice versa, or what?

An insurance company has 23 variations of a kind of life insurance. Are there 23 subclasses?

Is there a mapping between problem-space classification and design-space subclassing?

"Subclassing" optimizes lines of code; "Inheritance of interface" preserves a service set.

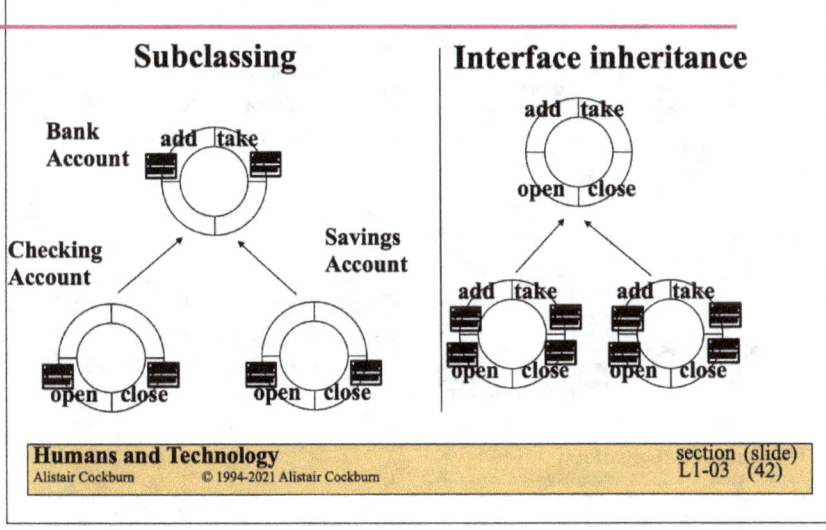

Design in Object Technology: Class of 1994

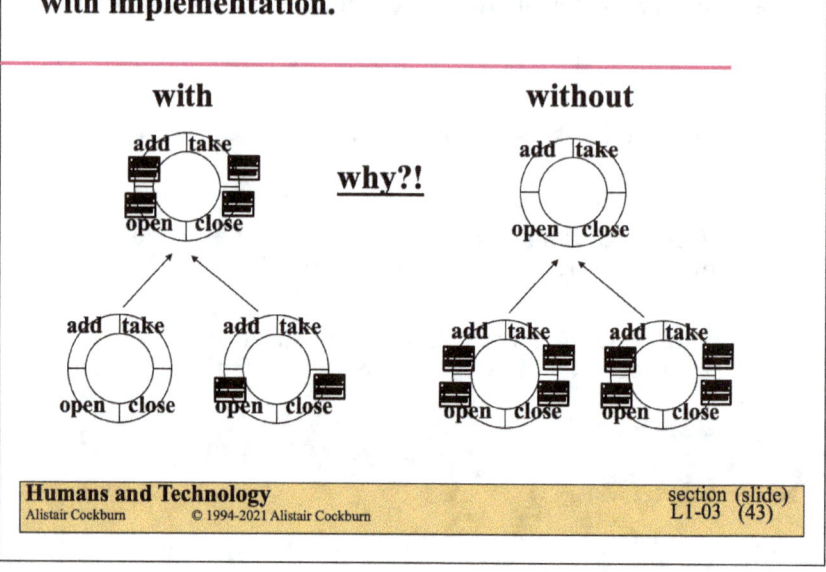

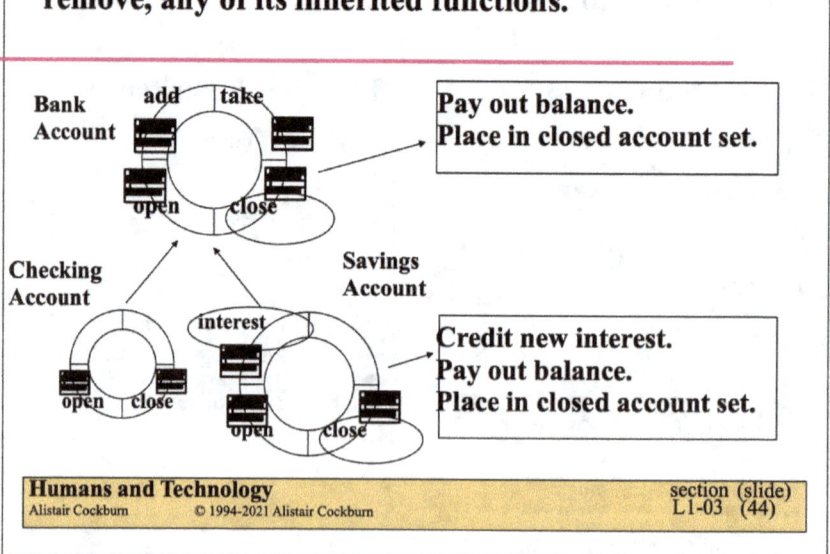

Design in Object Technology: Class of 1994

Inheritance is sensitive to implementation language and a class's final design.

1. **Smalltalk allows free polymorphism, so can call across class hierarchies.**
   C++ uses typed polymorphism, so can only call within a single class tree.
   - You may have to add classes repetitively.
2. **C++, Eiffel allow multiple inheritance.**
   - Good for building interface inheritance lattices.
   Smalltalk uses single inheritance.
   - But free polymorphism makes it less restrictive.
3. **Final implementation may use instance variables instead of subclassing.**

Rule of Design: Identify classes early, but leave inheritance to language-specific class design.

1. Identify which classes are needed.
2. Identify commonality between classes.
3. Let the class designer decide how best to implement the commonality
   - Interface inheritance
   - Subclassing
   - Instance data

Design in Object Technology: Class of 1994

Object Interaction Diagrams

Series on Object-Oriented Design

Interaction diagrams are crucial to understanding an OO design.

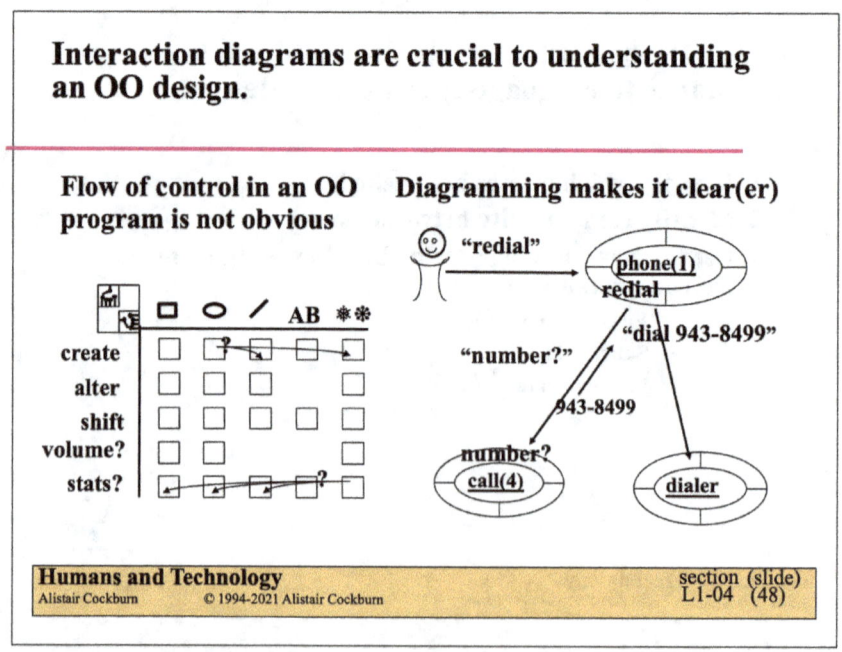

Flow of control in an OO program is not obvious

Diagramming makes it clear(er)

Design in Object Technology: Class of 1994

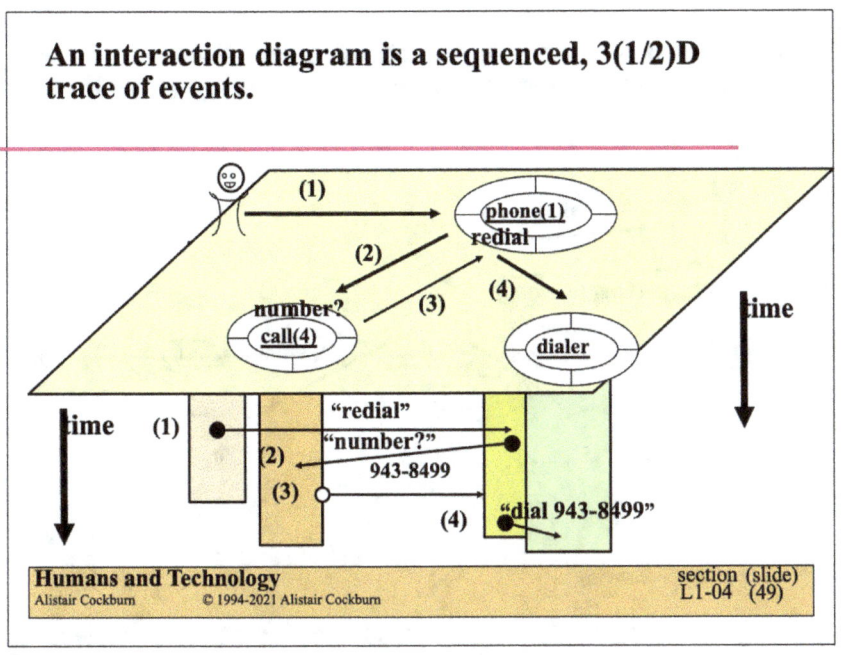

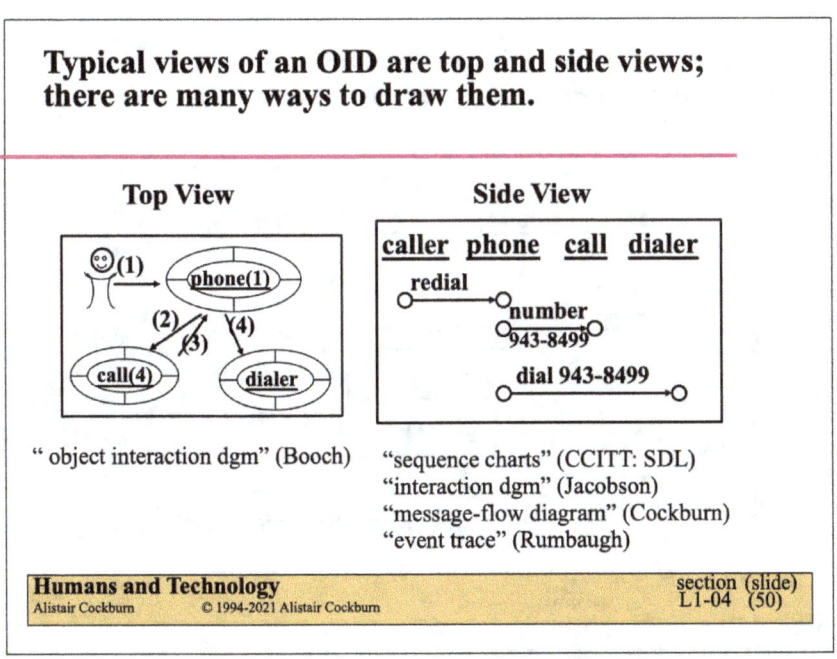

Design in Object Technology: Class of 1994

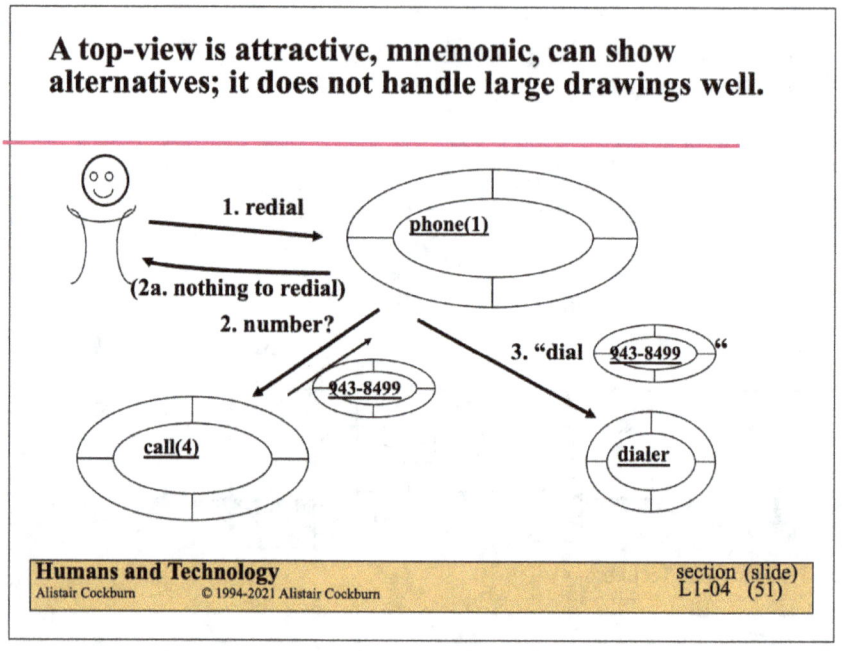

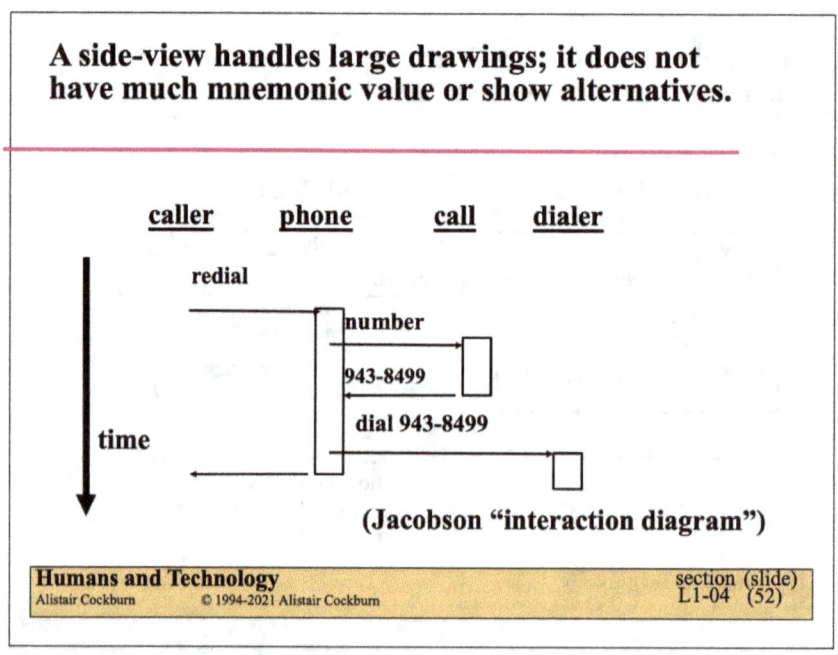

Design in Object Technology: Class of 1994

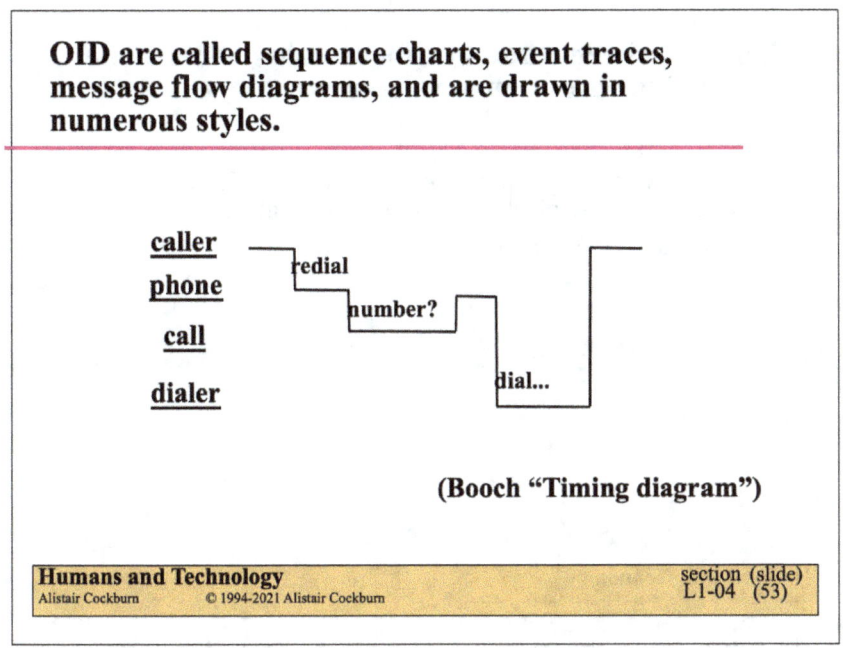

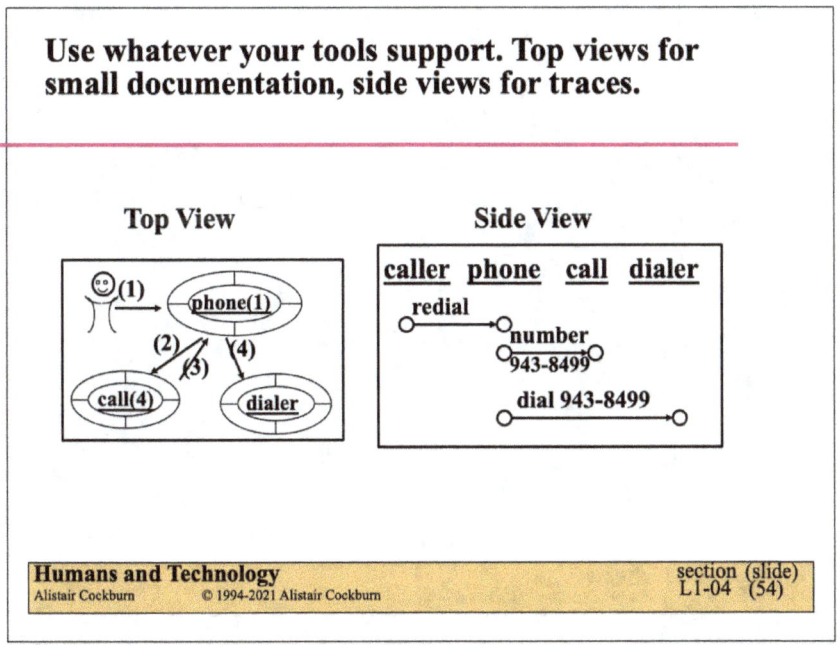

There have been studies on formalizing OIDs, none yet suited for program development

But, you can imagine some code generation from OIDs, and use of trace visualization tools...

...soon to arrive at a vendor near you.

Design in Object Technology: Class of 1994

"Responsibilities" in Object Design

Series
on
Object-Oriented
Design

Each unit in a system plays a role... it has a "responsibility" to the functioning of the system.

(This is always true, not just for Object-Orientation)

"Responsibility" was coined by Ward Cunningham and Kent Beck to describe essence of good OO designs.

<u>Naming and allocating</u> responsibilities is key to design.
-> determines system decomposition and flow.

An object can delegate to others, calling upon their responsibilities.

"Design with responsibilities" mirrors society,
So...
1. Pretend you were the object...
2. Ask yourself, "Is it really my responsibility to handle this?"
3. Ask, "Who would I call upon to help, who has the responsibility to help in that way?"
4. (Practice different scenarios to stress-test the allocation of responsibilities.)

A unit's role or responsibility can usually be stated in one or two short phrases.

"Knows about collecting money and giving change."
- Coffee machine coin/credit collection unit

"Knows its business purpose and mediates its business attributes."
- Generic statement for a business object.
- It does not claim to know how the data is stored.

"Knows how the data for a particular business object is stored."
- A "data broker" object for the business object.

"Knows and controls the details of a transaction."
- Transaction object (e.g. a Withdrawal)

Design in Object Technology: Class of 1994

The statement of responsibilities is the shortest description of a unit's requirements & function.

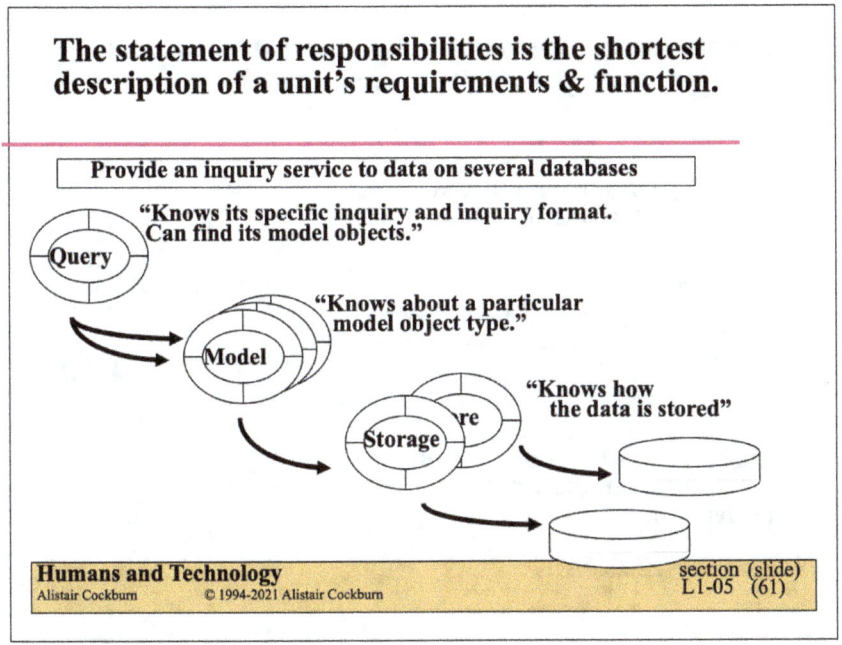

Provide an inquiry service to data on several databases

"Knows its specific inquiry and inquiry format. Can find its model objects." — Query

"Knows about a particular model object type." — Model

"Knows how the data is stored" — Storage

A main responsibility involves smaller responsibilities. Design to the main responsibility.

Responsibilities of an Order:
  "Responsible for its line items and its value.."

Sub-responsibilities:
- Knows how to add and remove line items.
- Knows the tax computation.
- Knows its final value,

Design <u>up</u> to the main one, or <u>down</u> from it, but capture the main responsibility, as it is the short-form summary of the object.

Design in Object Technology: Class of 1994

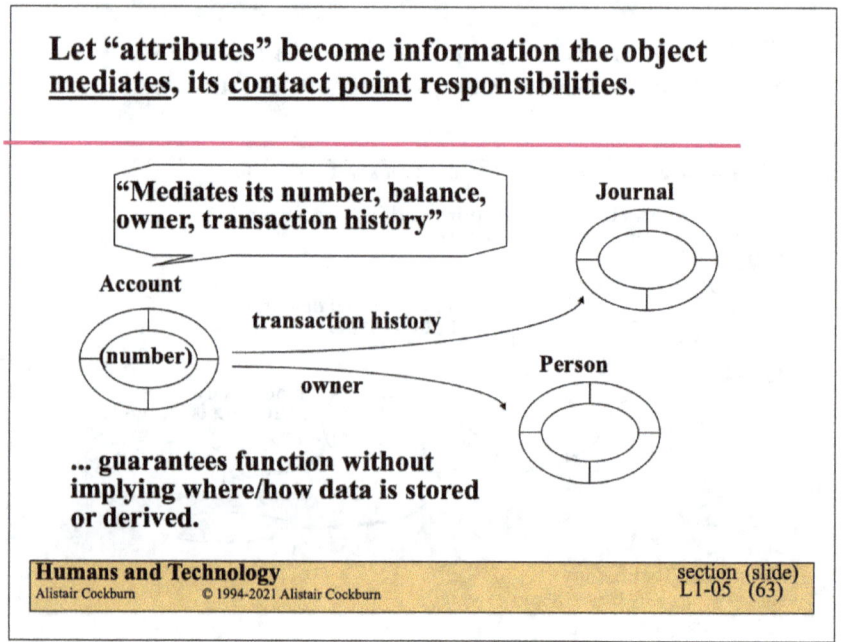

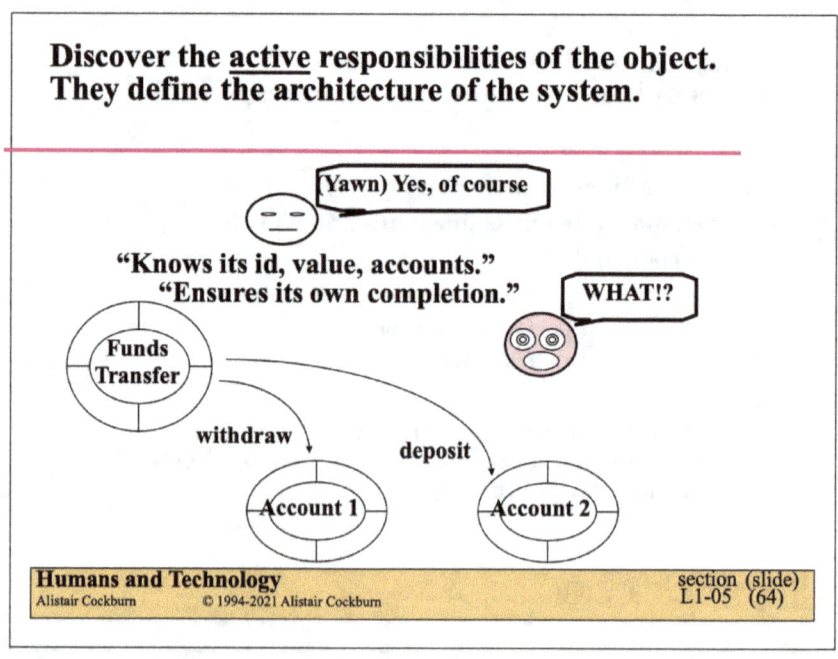

Design in Object Technology: Class of 1994

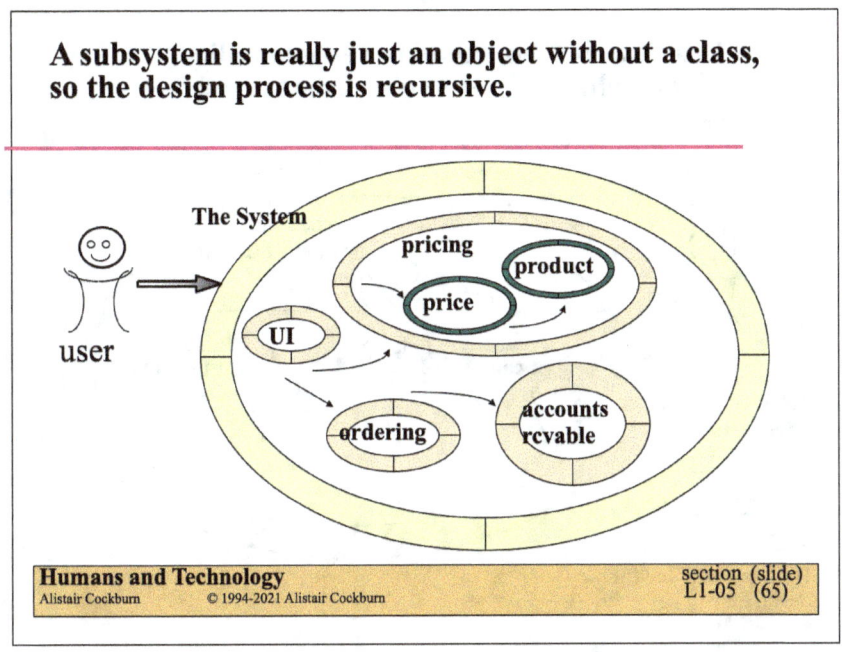

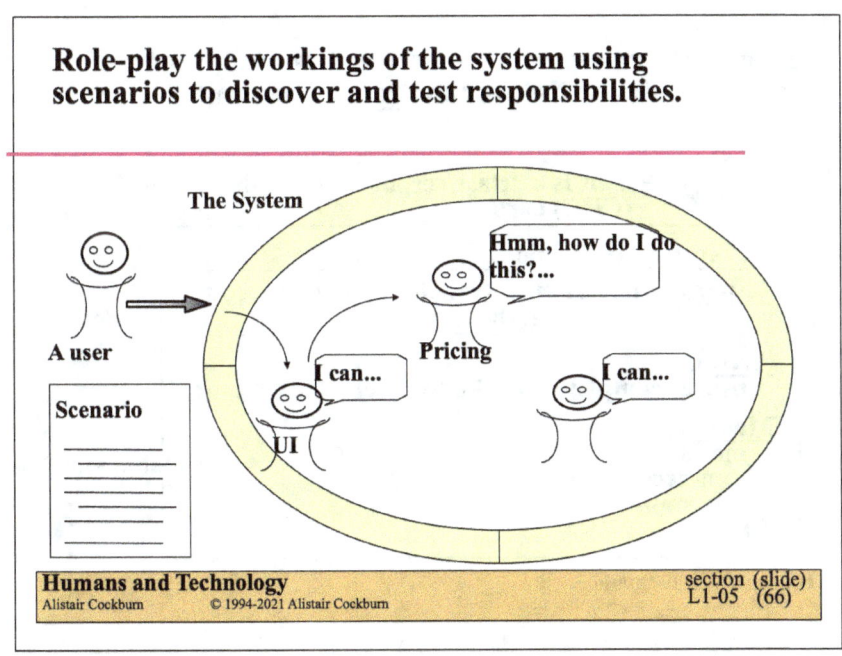

Design in Object Technology: Class of 1994

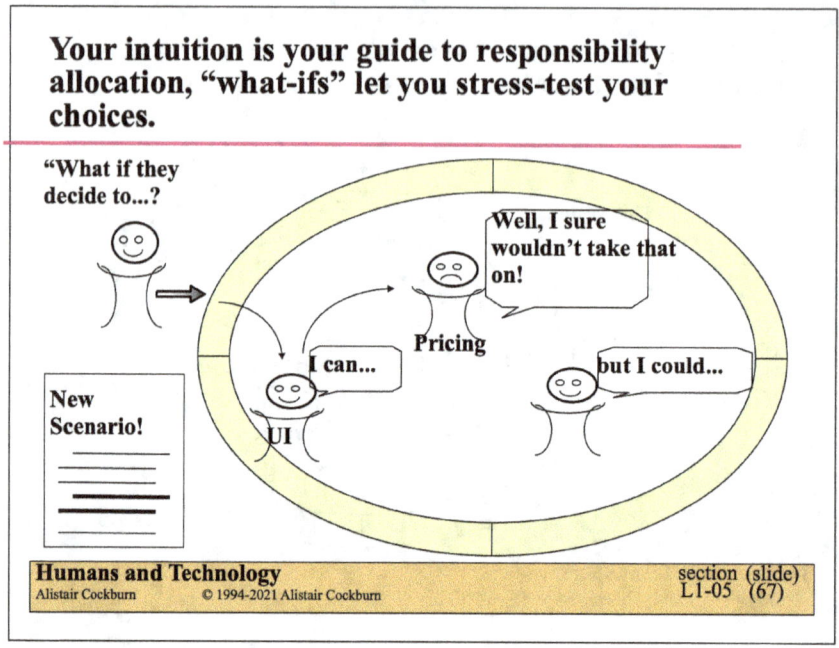

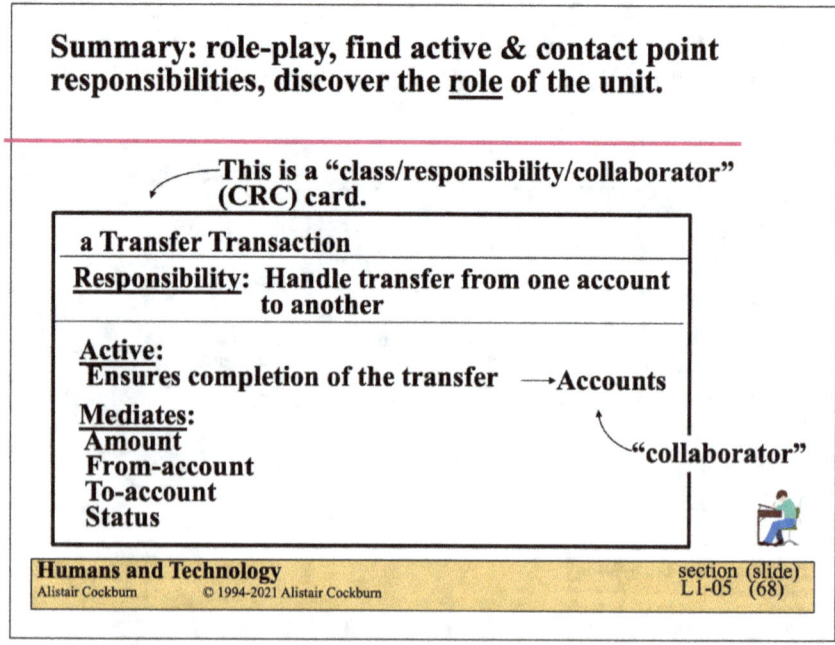

Design in Object Technology: Class of 1994

Using Scenarios in Design

Series on Object-Oriented Design

Humans and Technology
Alistair Cockburn © 1994-2021 Alistair Cockburn
section (slide)
L1-06 (69)

Usage scenarios provide the basis for design with responsibilities.

> **Responsibility-based design is based on role-playing a walkthrough of a scenario.**
> - This happens, then this happens,
> - Who takes care of that? ...
>
> **Multiple scenarios provide the basis for asserting,**
> - "This design delivers the required function."
>
> **Variation scenarios provide the basis for asserting,**
> - "The design responds thus to these likely variations."

Humans and Technology
Alistair Cockburn © 1994-2021 Alistair Cockburn
section (slide)
L   (70)

Design in Object Technology: Class of 1994

A scenario is a single trace of events with no branching. (It is therefore incomplete)

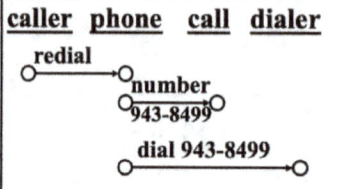

What happens if there is no last call to redial?

Scenarios start from a goal and given conditions, and go until the goal is achieved or abandoned.

Goal: Call them again, quickly, easily.
Conditions: Phone stores last phone number,
   There is a last call to redial.
Outcome: Last number is redialed.

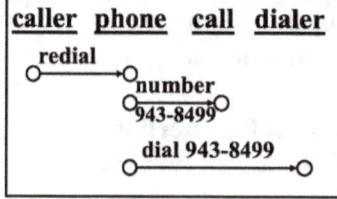

MFD

caller asks phone to redial.
phone locates number,
and has it redialed.

Text

Design in Object Technology: Class of 1994

A separate scenario is used if a different branch is taken.

Goal: Call them again, quickly, easily.
Conditions: Phone stores last phone number,
   There is NO last call to redial.
Outcome: Phone tell caller of failure and reason.

caller   phone   call   dialer
   redial
   o────────o
   "no last call to redial"
   o────────o

caller asks phone for redial.

phone finds no call to redial.
phone tells caller,
"no call to redial"

MFD   Text

A scenario is recursive - it can call upon other sets of scenarios

Scenario: claimant writes car insurance company requesting benefits from car accident.
Conditions: claimant eligible for benefits
Outcome: claimant paid benefits.

1. Claimant submits claim and details.
2. Insurance co. checks claimants file -- all ok.
3. Insurance co. assigns agent to <u>evaluate accident</u>
   • -- all accident details within policy outline.
4. Insurance co. <u>assesses value to pay.</u>
5. Insurance co. pays claimant determined value.

wow!

Design in Object Technology: Class of 1994

A set of scenarios helps establish and stress-test responsibility allocation.

What happens if the claimant is not even registered with this insurance company?

What happens if the other party is at fault but has no insurance? ... or is from out of the country?

What happens if the database is moved?

What if the insurance agent quits or dies?

What if the claimant dies?

What if we design it for 3270 and they decide to move to PCs ... or vice versa?

What if they go with small, mobile handsets?

Requirements scenarios treat a system under design as a single, black-box system.

Requirements scenarios explain and validate the functioning and purpose of the system.

They say What to provide, in a usage context, without saying How to provide it.

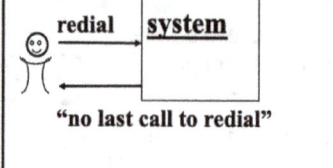

| OID | Text |

caller asks for redial.

system finds no call to redial.
system tells caller,
"no call to redial"

redial → system
"no last call to redial"

Design in Object Technology: Class of 1994

Requirement scenarios work at multiple levels of system and multiple levels of detail.

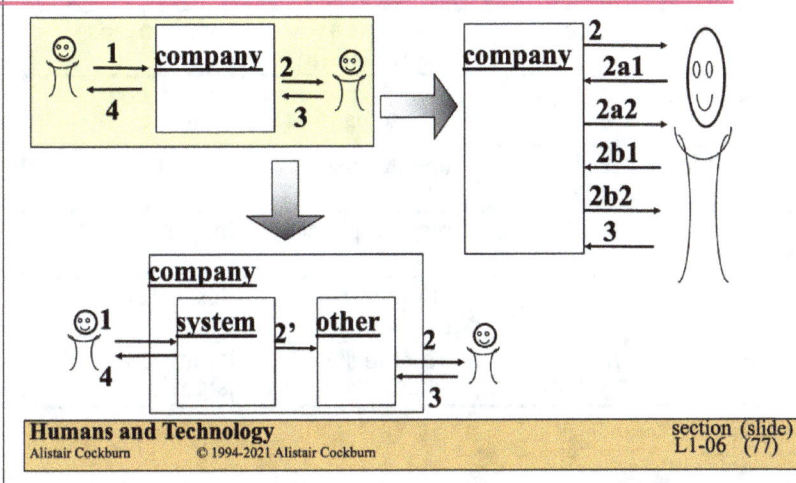

Humans and Technology
Alistair Cockburn © 1994-2021 Alistair Cockburn
section (slide) L1-06 (77)

Include scenarios of use for every user of the system -- include test and maintenance!!

Identify the actors who will use the system
   Actors may be people, computers, programs,
   end users, maintenance personnel, testers.
Make test and maintenance standard actors!

Humans and Technology
Alistair Cockburn © 1994-2021 Alistair Cockburn
section (slide) L1-06 (78)

Design in Object Technology: Class of 1994

Summarize functional requirements in 4-columns: actor, goal, system responsibility, data needs.

ATM actor	goal	system responsibility	data needs
account owner	withdraw money	give $, receipt; update balance	acct #, code, amount.
bank employee	refill cash drawer	update cash balance.	$ amount.
maintenance staff	refill paper	register paper (non)empty	paper present.
tester	test many situations	read/run test scripts, & produce report.	test scripts.
installer	initialize system	reset to start state.	"initialize" signal.

Humans and Technology
Alistair Cockburn © 1994-2021 Alistair Cockburn
section (slide) L1-06 (79)

Summary: write scenarios that are simple, black-box, staged in complexity, and complete.

1. Start from each user's goal, go to conclusion.
2. Fold sub-scenarios with same outcome together.
   Put sub-scenarios with different outcome into a different scenario.
4. Mention only systems that have been designed, leave closed what is yet to be designed.
5. Sets of scenarios are not always easy to structure, so use your imagination!
   - Sets of scenarios involve exceptions, parallel processing and asynchronous conditions!

Humans and Technology
Alistair Cockburn © 1994-2021 Alistair Cockburn
section (slide) L1-06 (80)

Design in Object Technology: Class of 1994

Use cases and goal statements

Series on Object-Oriented Design

A "use case" is a group of requirements scenarios with two outcomes: goal attained or abandoned.

Describes usage of the system - context for the "why".
"Why are we requiring this function?"
Shows the intention of a key user: their goal.
System may be able to deliver goal or not (conditions)
Use case covers
  0. goal delivered without problem,
  1. goal delivered after failure recovery,
  2. goal abandoned.

Design in Object Technology: Class of 1994

A use case is characterized by actor and goal. It contains scenarios.

An actor is a person, computer or other active thing.
"What is the actor trying to accomplish?"
"What scenarios can occur in pursuing this goal?"
   Both success and failure scenarios.
Examples:
   Operator wants to find desired music selection.
   Owner wants to withdraw money from bank.
   Customer wants to trade money for goods.
   Customer wants to register change of address.

Humans and Technology
Alistair Cockburn   © 1994-2021 Alistair Cockburn
section (slide) L1-07 (83)

A use case has success and failure outcomes, a scenario has only one outcome.

Primary Actor: Account owner Primary Actor's Goal: withdraw money			Use case characteristic information
Knows codes, Has funds.	Knows codes, No funds here, Other funds.	Knows codes, Not sufficient funds anywhere	Scenario conditions
Presses button.	Presses button.	Presses button.	Trigger
Get codes. ok. Ask amount.ok. Give money.	Get codes. ok. Ask amount.Nok. Transfer funds ok. Give money	Get codes. ok. Ask amount. Nok. Transfer funds. Fail. Refuse money.	Scenario steps
Succeed		Failure	Outcome

Humans and Technology
Alistair Cockburn   © 1994-2021 Alistair Cockburn
section (slide) L1-07 (84)

Design in Object Technology: Class of 1994

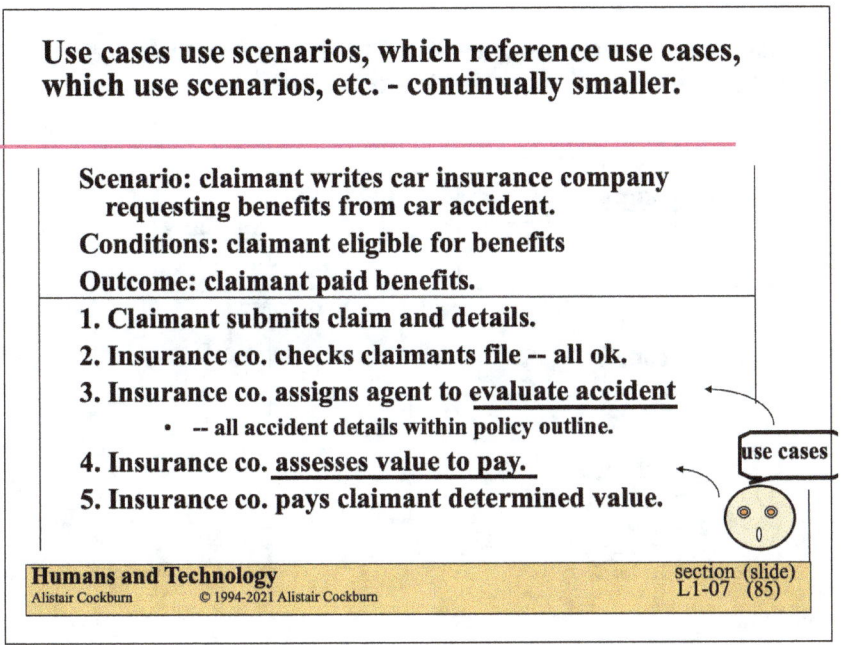

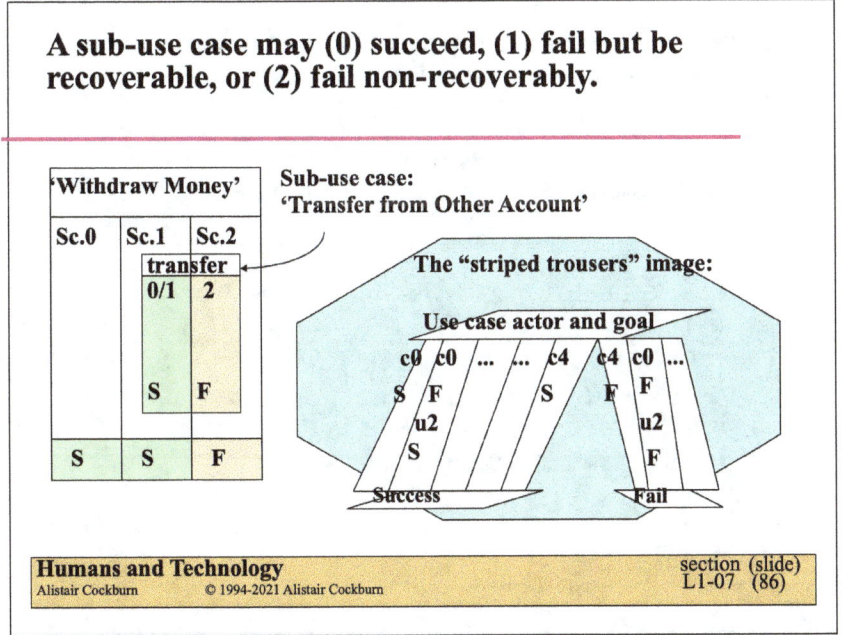

Design in Object Technology: Class of 1994

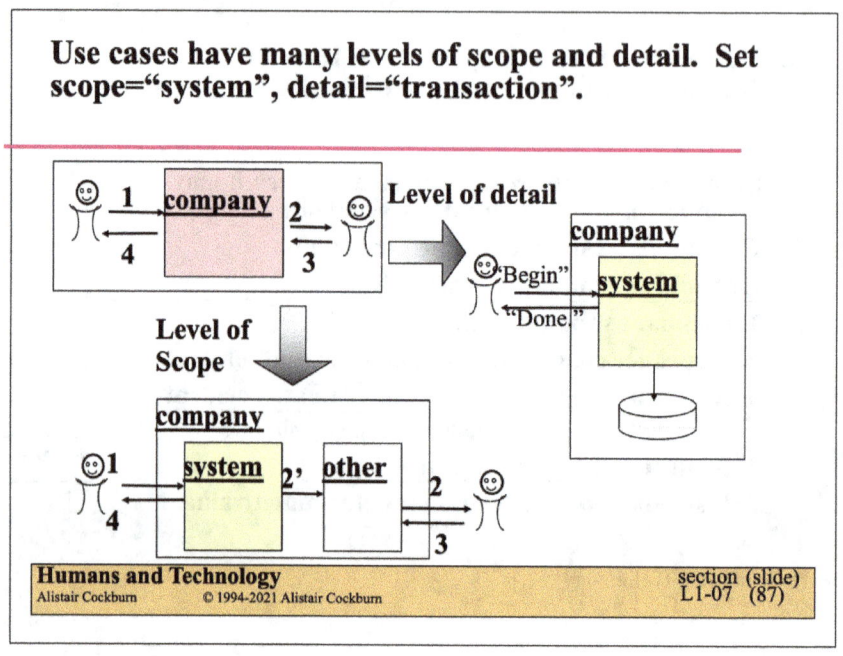

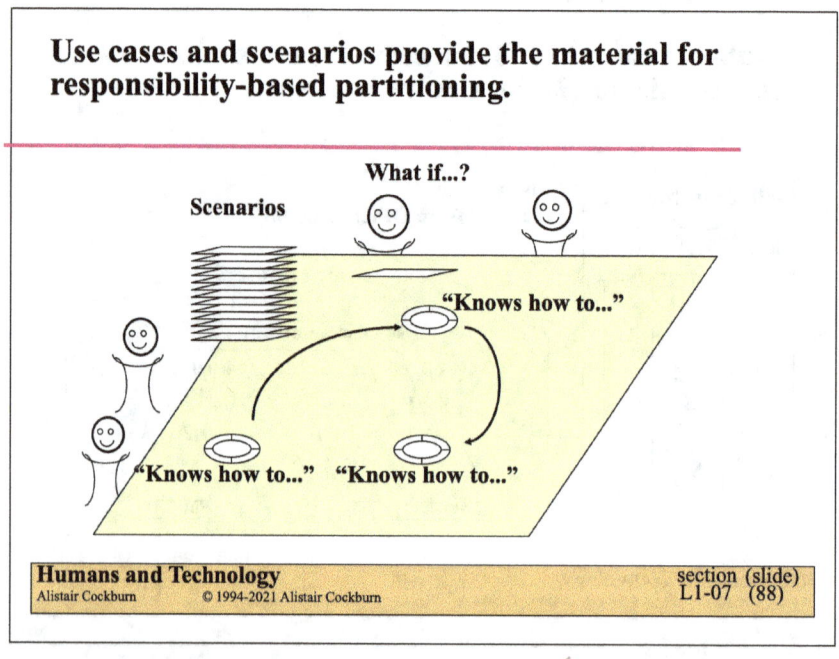

Design in Object Technology: Class of 1994

Object interaction diagrams show use cases for subsystems.

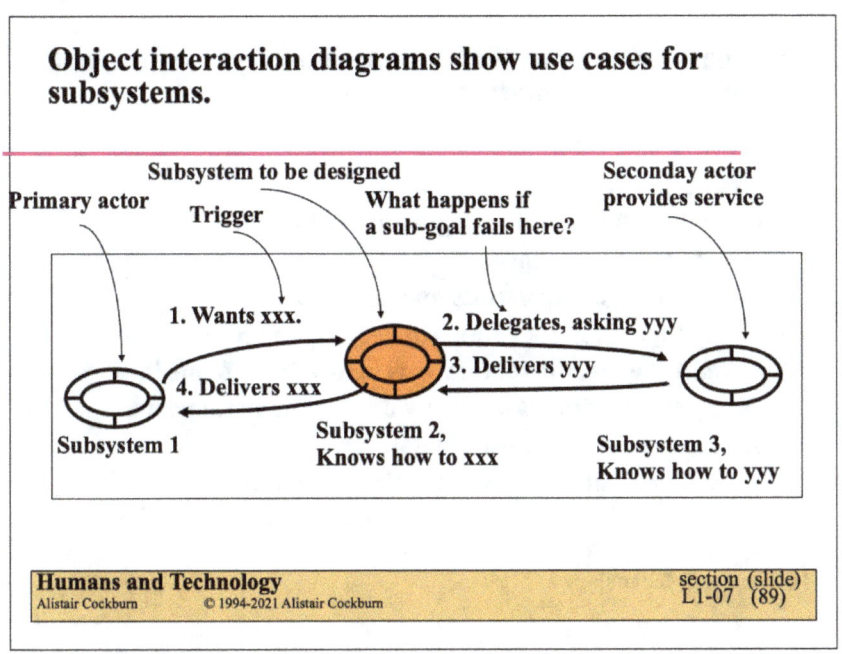

Summarize interface requirements in a table: use case, trigger, secondary actors, interface types.

use case	trigger	secondary actors	interface type
withdraw money	key press	bank central, cash dispenser	database table. hot link.
refill cash drawer	lift $ sensor	cash dispenser	hot link
test many situations	program command	server computer	flat file.
initialize system	key press	none	---

Design in Object Technology: Class of 1994

Use cases & responsibilities only handle function, not business rules & math.

Request function,
Show how function is delegated/partitioned.
But, how do you show
- @ Every order gets its own invoice.
- @ Final value = ValuesTotaled * tax percent.
- @ Whenever inventory gets below 25%, triple it.
- @ Allow withdrawals only during the fiscal year.

Append these to the use case(s).
    Ok to create annotations ("just a tool problem").

Summary: Actor / goal / success & failure scenarios. System scope, transaction level of detail.

1. The primary actors, their goals.
2. Hunt for the failure scenarios.
   - The "main" success scenario reaches the goal easiest.
   - Note conditions, outcome, trigger, steps.
3. Find end-customer and system-level scopes.
4. Find business transaction level of detail.
   Sub-usecases can go below business transaction level for utility.
5. Use cases use scenarios use use cases.
6. Respsonsibility-based design needs scenarios.

Design in Object Technology: Class of 1994

Models, shields and algorithms

Series on Object-Oriented Design

The "model" (class) mirrors the world. A shield (class or service) protects against design changes.

Goal: Reduce the "trajectory of change"
1. "Mirror the world in objects (nouns)"
   It is easy to nominate hundreds of model classes.
   e.g.: a Bill, a Billing Line Item, a Product, a Price
2. Decide which to keep using responsibilities.
3. Look for Shields in services and classes:
   "What if we have to change xxxx to yyyy?"
   "We can't yet decide how ..."
   e.g.: an accessor, a Strategy, a View, a Librarian

Design in Object Technology: Class of 1994

A shield (service or class) provides an interface to allow variation.

Consider known "change cases" (John Bennett):
  "We know this will change..."
  "We don't know how it will be stored."
  "We expect it eventually to be computed on the fly."
  "Let's hard-code it in the prototype"
Shields protect your clients, reduce trajectory of change.
Scenarios / responsibilities help you find shields.
  - A service lets you change <u>the</u> implementation
  - A class lets you subclass for many implementations.

Introduce shield classes sparingly.

A shield class introduces complexity, delay.
e.g.: View classes as shields:
  "External representations change often"...so
  Separate the View from the Model
    - Reuse same model using different views.
  Now have View classes and Model classes.
    - more complex but smaller changes
Resist temptation to introduce a "shield class" to avoid thinking!

Design in Object Technology: Class of 1994

The best design makes Model objects Shields!

Good Shields ("Promotion") look like Models
   plus protect lots of future designs.
Look for generic business artifacts
After agonizing scrutiny and trial:
   Video Rental co. has Promotions for Customers
   - What is the responsibility of a Video?
   - of a Promotion?
   - of a Movie?
   - What kinds of Promotions?

Algorithms are key to speed, are forgotten by Responsibility design.

Responsibilities, models, shields address partitioning, protection for design changes.
Applications need SPEED.
   Tune code - 10% - 30% improvement
   Change algorithm - 300% - $10^{**}n$ improvement
Algorithm design is different from partitioning.
   Design algorithm first
   Partition to protect design variations.

Design in Object Technology: Class of 1994

Improve data structures and algorithm exponents.

Data structures hide their algorithmic complexity
  Linked list = O(n)   Array = O(1)
  Tables = O(n**2)   Naive graphs = O(n**3)
Program FIRST for design safety,
Measure LATER for bottlenecks
Change exponent at the bottlenecks.

Summary: Start with model classes, find shields to protect design choices, algorithms to get speed.

1. Listen carefully for the concepts in your conversations.
   The concepts become Model classes.
2. Stress test the responsibilities with variations.
   Place a Shield where you will create change.
   Look for subclass needs.
3. Improve program speed by improving algorithms (exponent improvements).

Design in Object Technology: Class of 1994

Recursive Design

Series on Object-Oriented Design

Communicating systems are recursive in nature, from enterprise down to program modules.

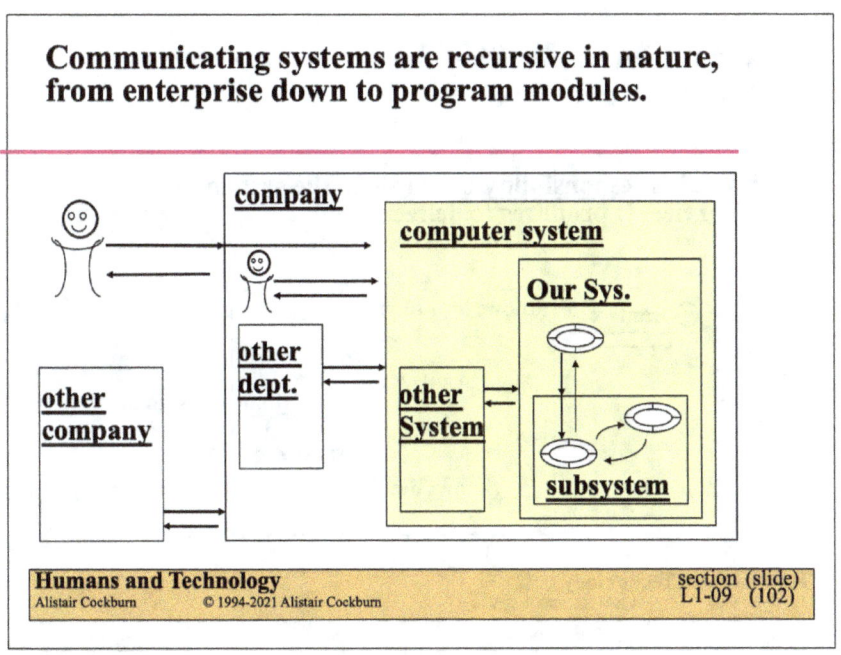

Design in Object Technology: Class of 1994

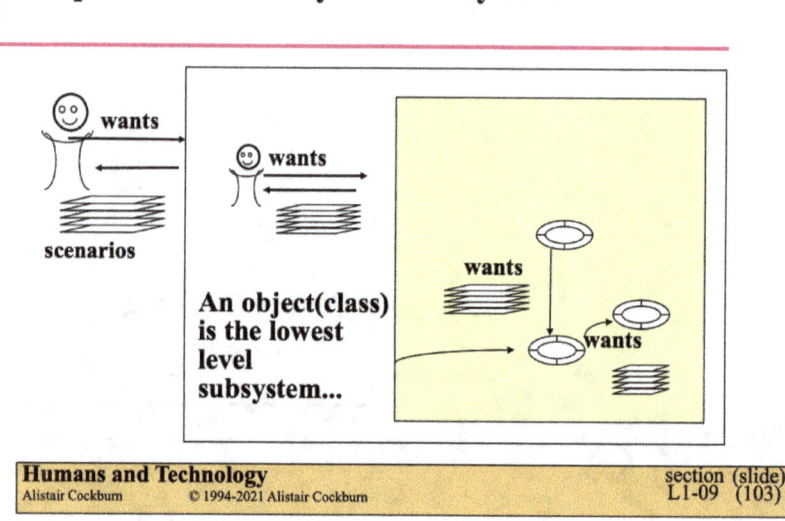

Use cases (collected scenarios) specify service requirements for a system of any size.

An object(class) is the lowest level subsystem...

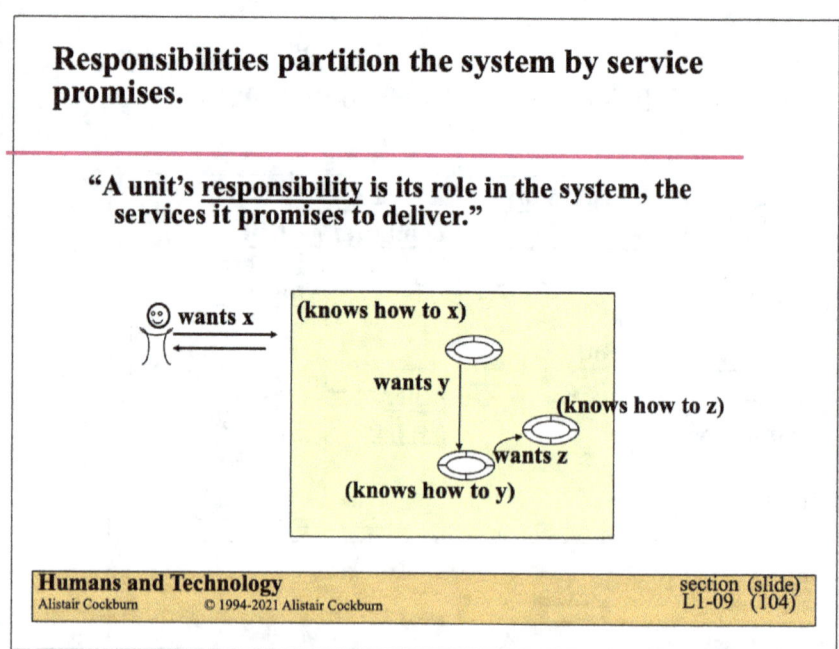

Responsibilities partition the system by service promises.

"A unit's <u>responsibility</u> is its role in the system, the services it promises to deliver."

Design in Object Technology: Class of 1994

Interaction diagrams show how the sub-services combine to deliver the system's promised service.

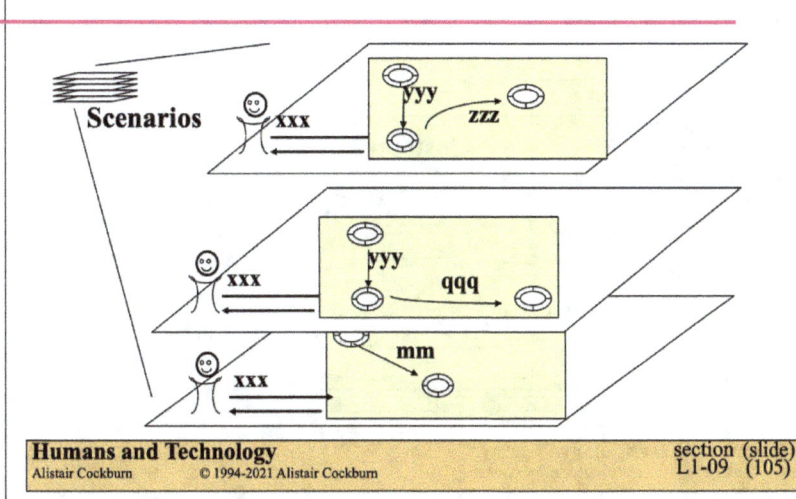

The interaction diagrams act as requirements scenarios and partial designs on the subsystems.

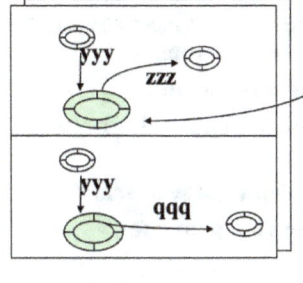

Design this to deliver yyy.
It sends zzz or qqq to do so.

A single class instance can be described by its services and responsibilities

A "framework" is a system of multiple objects, needing clarification through diagrams.

Design in Object Technology: Class of 1994

Each level of design makes sense by itself. One level of design is a "business" level.

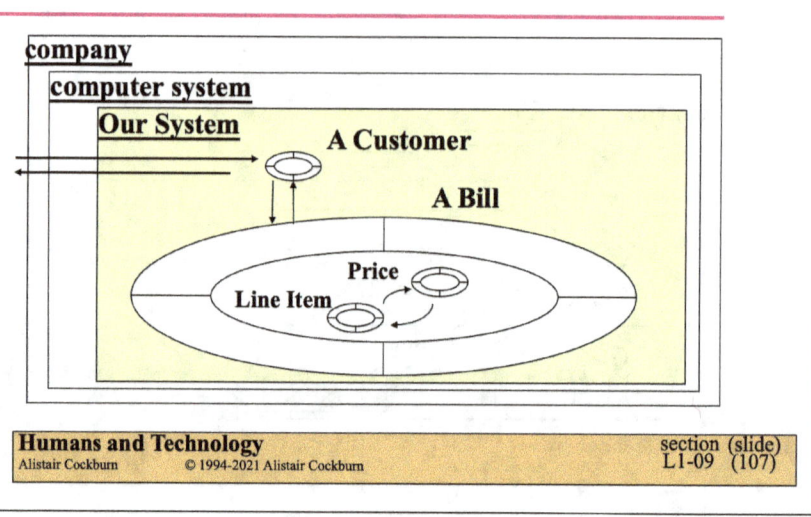

Summary: Use case & responsibility based design works from system down to class design.

1. Use cases collect scenarios that treat a service-providing system as a black box (functional requirements).
2. Responsibilities partition by service promises.
3. Interaction diagrams show service dependencies between the subsystems.
4. Interaction diagrams act as requirements scenarios.
5. Design the subsystem using the service promises.
   Down to the level of classes.

As soon as possible, eliminate sequential requirements,
   Use pure service statements.

Design in Object Technology: Class of 1994

Incremental Development

Series
on
Object-Oriented
Design

Incremental development is piecewise development and integration of the system.

- (Do not try to specify it all first, analyze it all, design it all, then test it all.)
- Specify as much as comfortable, design and build a piece. Learn from that experience.
- Design and build what can be kept in a person's head. Learn from each segment.
- With better knowledge, design the next part. Integrate. With better knowledge, ..., ...

  (This is OLD news. pre-1980)

Iterative development is learning from experience, and going back to a segment to improve it.

(Do NOT assume you will do your best work first. Estimate where you will come up short.)

Build, incrementally. Test, evaluate, measure. Decide the value of improvement.

Plan to make changes. You will anyway.

    Allow 10-20% schedule time for improvements.

(This is more old news. 1980's)

Managing to incremental and iterative development is <u>hard</u>.

Reasonable worries with no easy answers:
- "This is a fixed-price, fixed-time contract. Enter your bids."
- "How many people of what skills do we need to hire next year?"

Less reasonable, but common phrasings:
- "What do you mean, you don't know what you are going to build?"
- "What do you mean, you plan to build it wrong the first time !?"

Design in Object Technology: Class of 1994

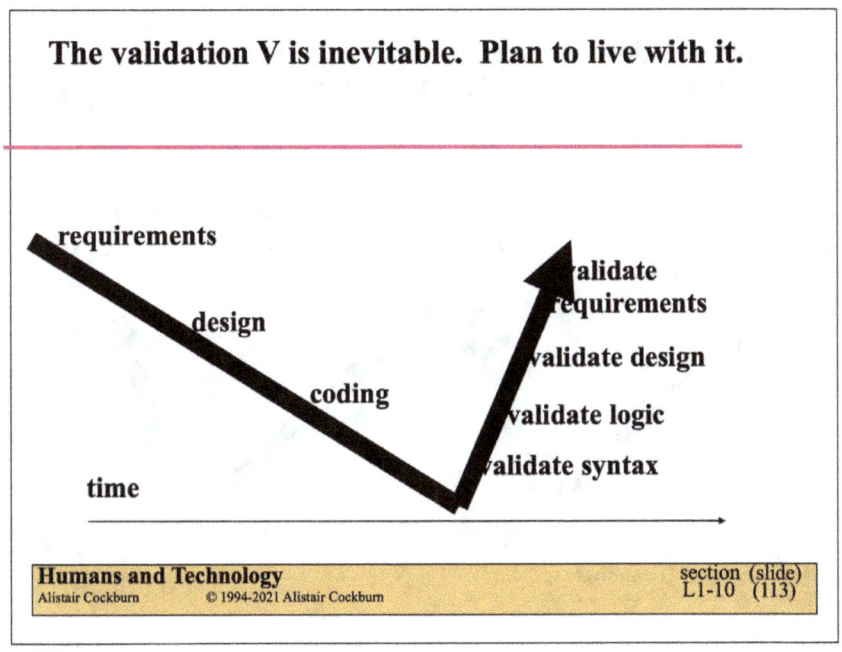

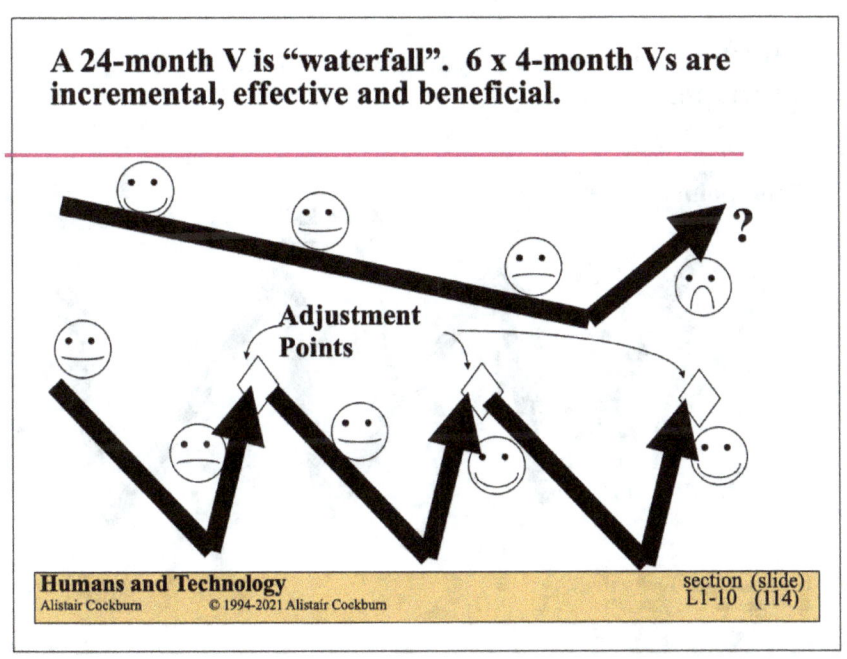

Design in Object Technology: Class of 1994

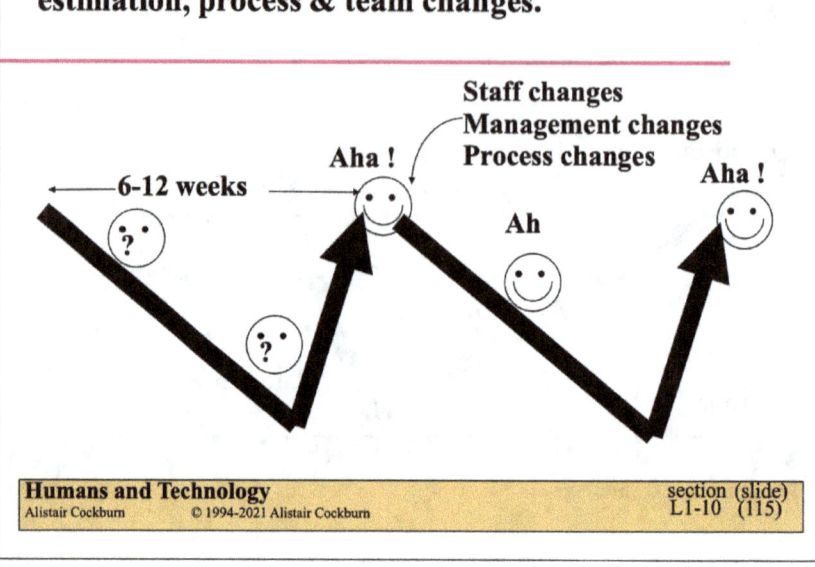

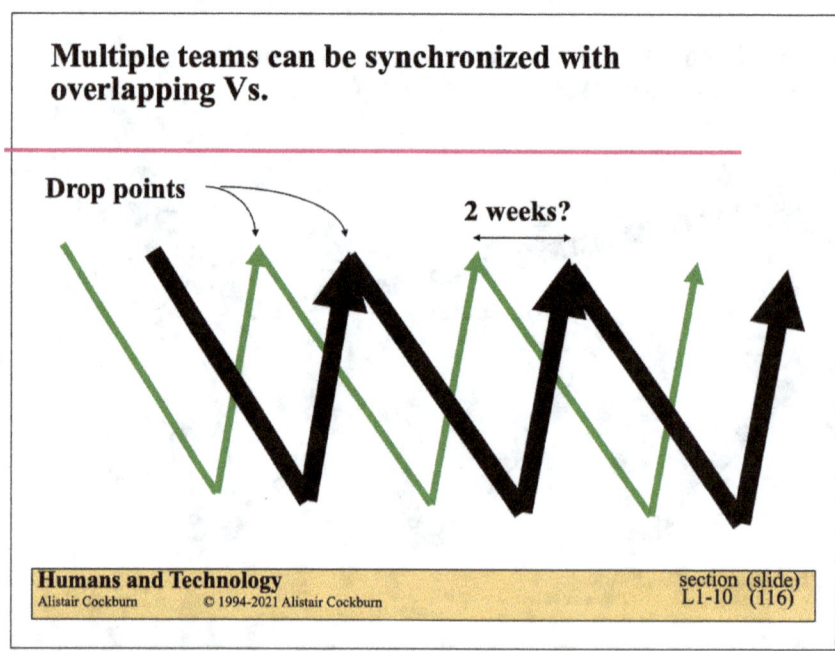

Design in Object Technology: Class of 1994

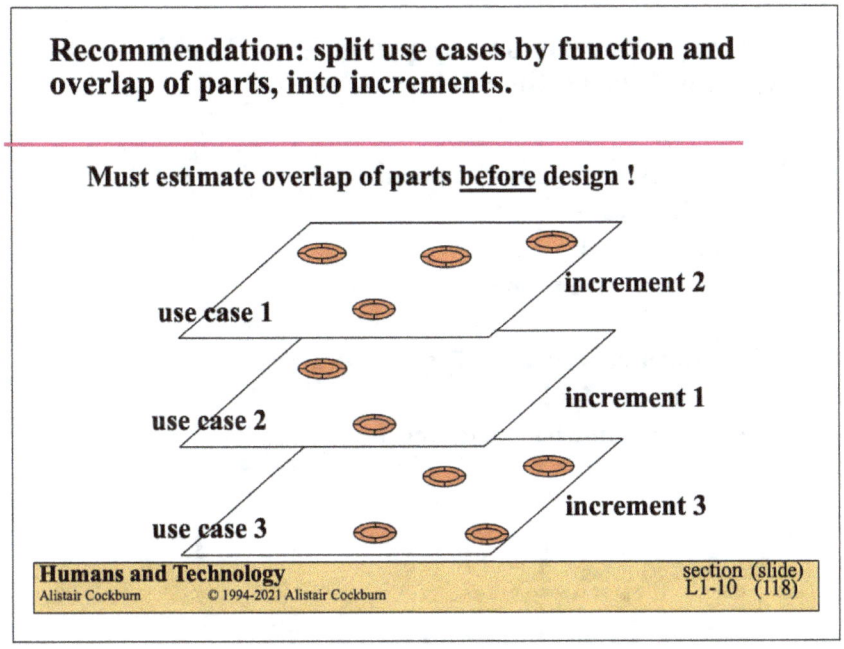

Design in Object Technology: Class of 1994

Iterative development is even harder than incremental development.

Iterative development requires recognition, planning and control of your weak areas!
1. Use shield classes and shield subsystems
   - e.g. separate UI from Model
2. Guess where understanding is weak / where change is likely -- and what it will take to fix.
   - e.g. new algorithms, new architectures, user interface
3. Plan 2-3 months for improvements at end
   - Performance, quality, simplicity
4. Note and track your risks and exposures.

Humans and Technology
Alistair Cockburn   © 1994-2021 Alistair Cockburn
section (slide) L1-10 (119)

Summary: Develop one stage and learn from it. Regroup, develop the next stage better.

Incremental development enhances
  estimation
  productivity
  process improvement
  morale
Split functionality into a <u>first</u> usable piece.
Work in (e.g.) 6-12 week units.
Learn each time how to do the next better.

Humans and Technology
Alistair Cockburn   © 1994-2021 Alistair Cockburn
section (slide) L1-10 (120)

Design in Object Technology: Class of 1994

Documentation

"A program reflects your current understanding. As your understanding changes, so does the program."
— Ward Cunningham

Series on Object-Oriented Design

Ideally, documentation leads someone else to understand how you came to make decisions.

- ... but we do not know how to do this, and so we work with "design notations".
- Design notations show the result of the decisions, not the decisions or the rationales.
  - Peter Naur (of Algol fame) considers programming as "theory-building"!
- A program's source code and a CASE tool drawing are both design notations.
  - Consider them two languages for one purpose. A reason for CASE failure is having only the result of the decision, but creating two, incompatible, design notations.

Design in Object Technology: Class of 1994

Use cases + responsibilities + interaction diagrams show context and result of decisions.

Use cases and scenarios are attractive because they show the context of the requirements, how a requirement is used.

So, a requested change can be evaluated in the context of the system's use.

After the decisions have been made, interaction diagrams show how the pieces work together.
- (And therefore may best be generated from the running system!)

In this course we document context, structure, responsibilities and interactions.

ER diagrams show business requirements, but lack context.

They show the result of a decision, but not the context, the rationale, or the validation.

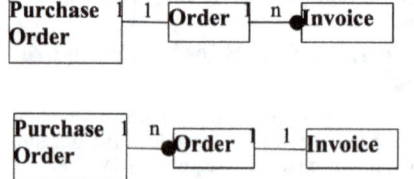

Which is correct? Why?
When can you challenge or change it?

Design in Object Technology: Class of 1994

Open issue: how to capture design rationales, taken decisions, open decisions.

There are too many decisions for each to be written.
   Peak rate may be 10 decisions in one minute.
   Which ones do you document?
You want to describe:
   1. The decisions that have been taken.
   2. The decisions or variations still open.
   3. What rules any variations must follow.
There is ongoing search for workable ideas:
   - MacApp and Taligent frameworks
   - Extensible games, design patterns

"Design patterns" are a fresh look at how one might document designs.

Gamma, Helm, Johnson, Vlissides, "Design Patterns"
The design patterns group is taking a new look at how to document designs, trying to address:
   1. The context of the requirements.
   2. Which decision points were taken.
   3. Which variations are compatible.
   4. How the current design works.
...these problems show up spectacularly in frameworks!

Design in Object Technology: Class of 1994

The design patterns catalog is trying a fresh approach to documentation.

1. State the problem the pattern addresses (text).
2. State general shape of the solution (ER diagram).
3. Describe how the solution works (text + dgms).
4. Name related patterns.

Ideas based on Christopher Alexander's work on documenting buildings.

Design pattern 1: Composites*

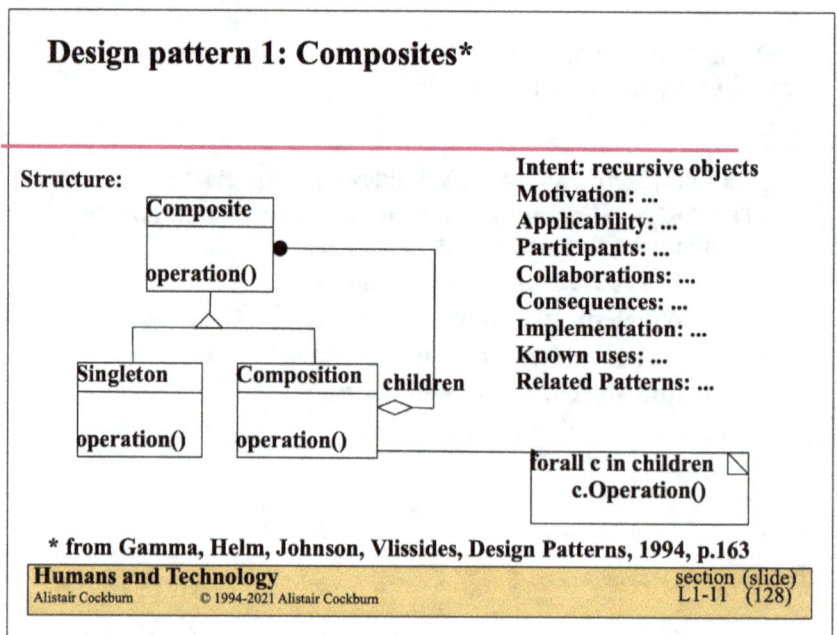

* from Gamma, Helm, Johnson, Vlissides, Design Patterns, 1994, p.163

Design pattern 2: Bridge*

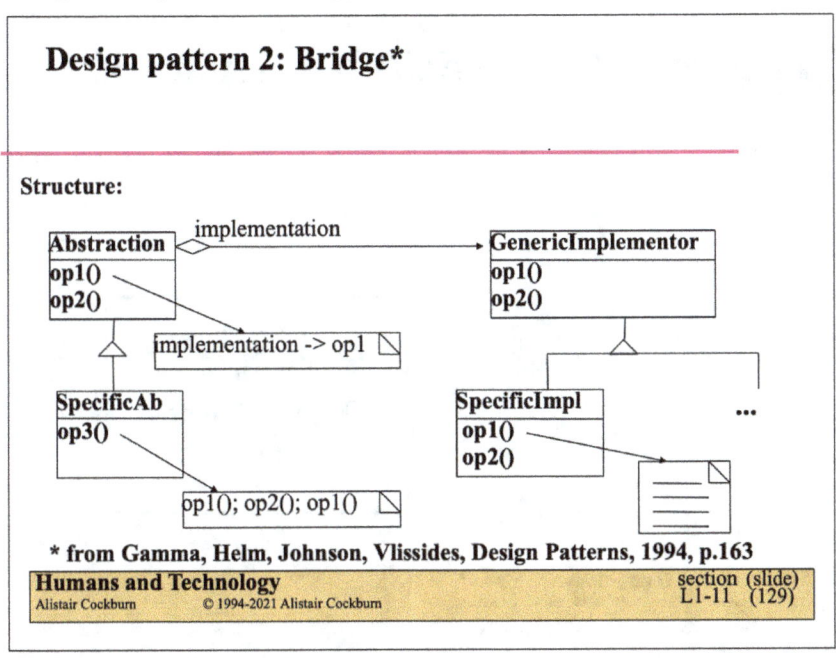

* from Gamma, Helm, Johnson, Vlissides, Design Patterns, 1994, p.163

Summary: Show how the system works in context, plus some reasoning and variations.

1. Use interaction diagrams (text or graphics) to describe the context of operation.
2. Use responsibility statements to succinctly describe the partitioning scheme.
3. Describe (text) reasoning behind non-obvious partitioning decisions.
4. Use ER, FSM, or math in amounts as needed, for business constraints or key ideas.
5. Mark shield objects with the planned variations.
   e.g. computed vs. stored vs. constant values.

Design in Object Technology: Class of 1994

Frameworks; Model/View Separation

Series on Object-Oriented Design

A framework is a complete or partial design that can have sections replaced or varied.

A framework can be any size, from 2 classes to a full-blown application.
- TWA frequent flyer
  - non-OO, IEF model, full application
  - traded between 3 airlines
- Model-View-Controller
  - 1976 Trygve Reenskaug Smalltalk-80
- Taligent and SOM persistence frameworks

Frameworks incorporate a lot of design knowledge
- Costly to build
- Great savings in use

Design in Object Technology: Class of 1994

Metaphor: a framework is an alterable game.

1. When completed, it plays a multi-party game.
2. It provides a structure that invokes the pieces.
3. Some rules are fixed, some can be varied.
4. New owner can vary and substitute parts.
5. New owner needs to know:
    - what game it plays
    - what can be varied and what not
    - how the variable areas work together.

Model-View-Controller is a commonly used OO framework.

Problem: Objects' appearances at the UI are changed far more often than their contents.

Cost: Any change to an object may damage it.

Solution: Separate the appearance from the heart. Make 2 or even 3 objects.

How the game plays: The trigger signal hit the controller, who decides whether to wake the model. The view registers itself with the model as interested in changes. When the model changes, it notifies any view registered with it.

Benefits: multiple views, no damage to model.

Design in Object Technology: Class of 1994

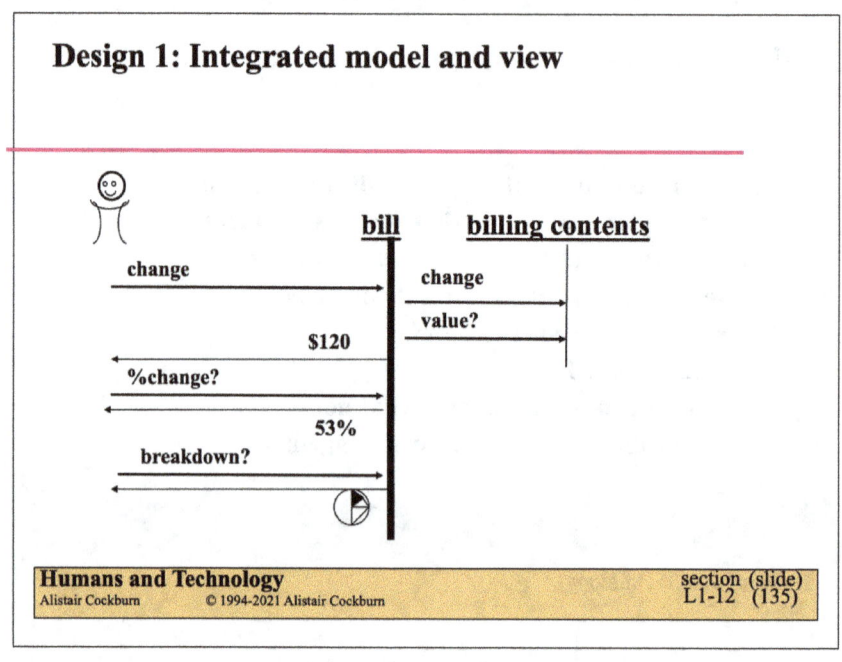

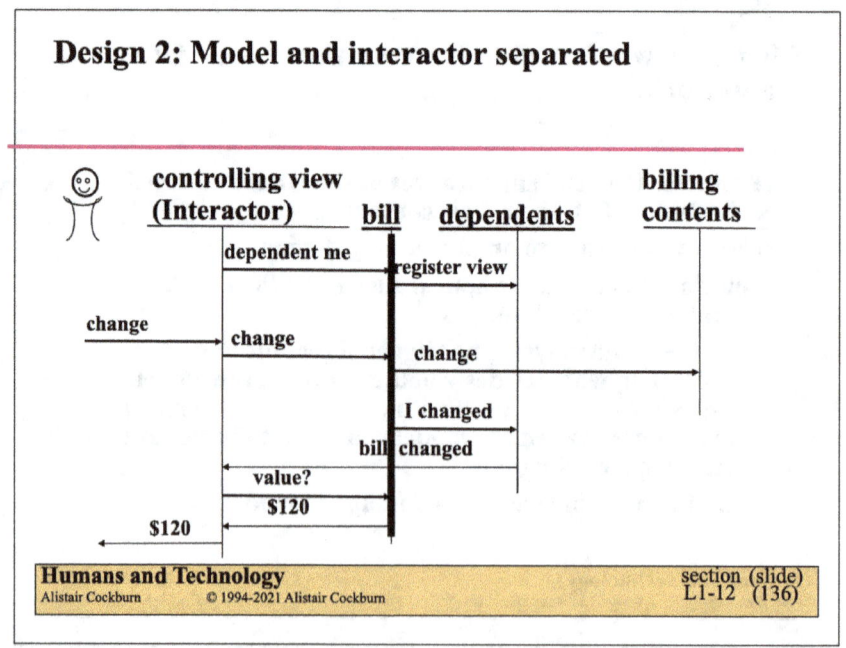

Design in Object Technology: Class of 1994

Design 2: Model and multiple interactors

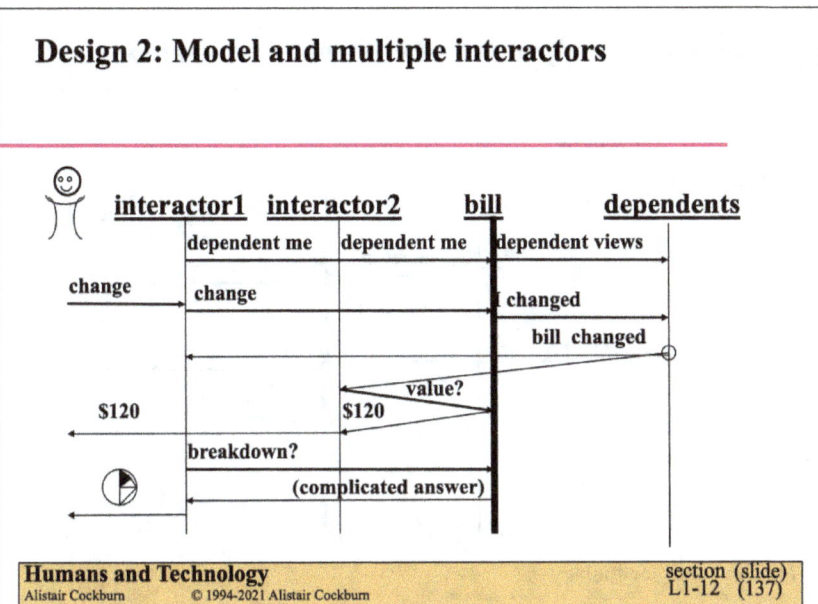

Design 2: Designer needs to know Dependents, Changed, and the model's public interface

"Any new view must send a 'dependent' message to the model with itself as argument.

Views will be sent a 'changed' message having model as argument, whenever model changes value.

View must know the public interface of model.

View has responsibility to render the object.

Model has responsibility to stay consistent and notify views of changes."

- Alternatively, view might send an "interested in:" message with the value(s) it want to be told about.

Design in Object Technology: Class of 1994

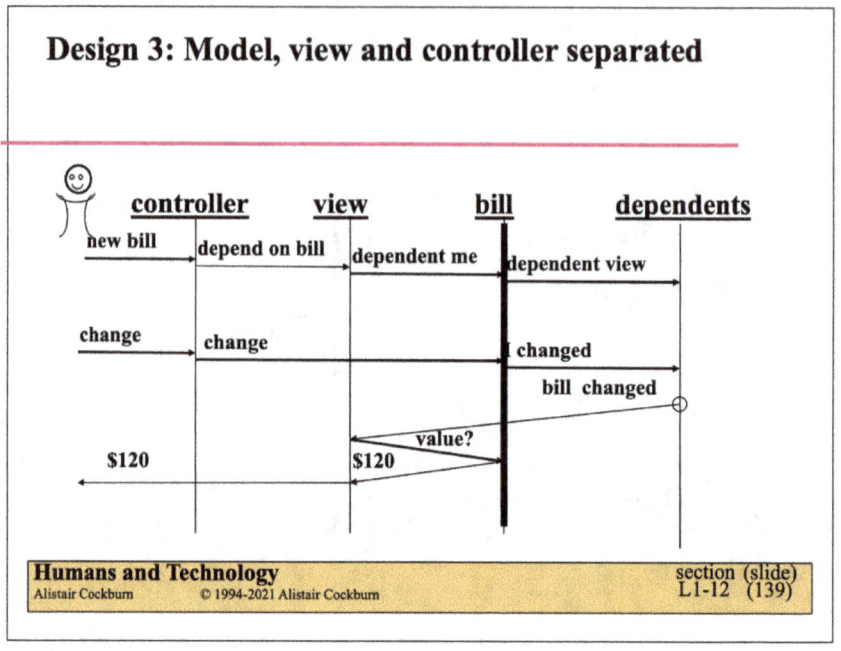

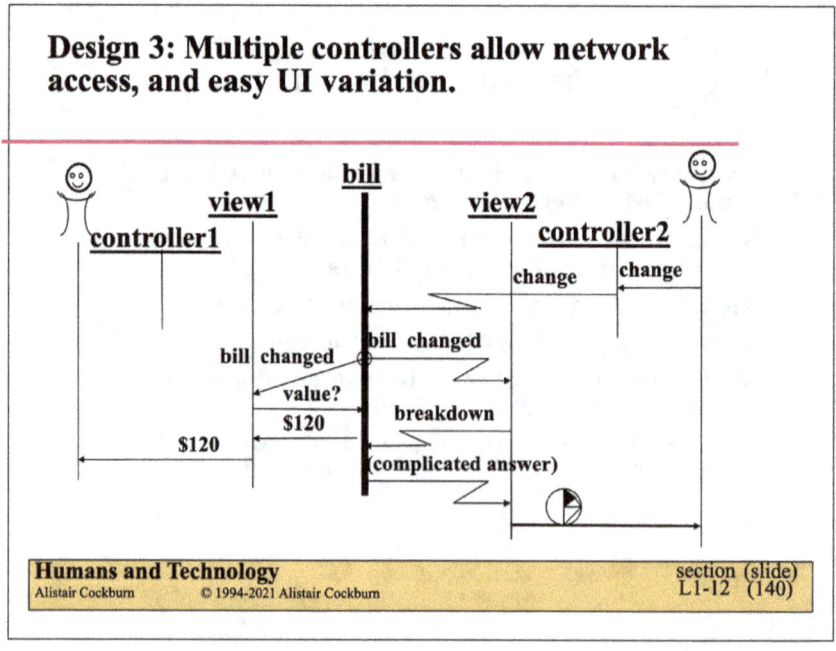

Design in Object Technology: Class of 1994

Any frequently solved problem can become a framework.

- Persistent data, asset management, transaction histories, user interfaces, network conversation, even a sorted collection.
- Only the non-competitive portions of a framework will reach the open market.
- Taligent, Template Software, IBM (SOM) are coming to market with libraries of OO frameworks.
- Experts suggest the best way to design a framework is to design the same thing several times under different circumstances.
  - Only then do the issues become evident.

Summary: Expect frameworks to become standard, but hard to design and understand.

- Typically, several designs are done before the tunable areas become apparent.
- Documenting frameworks is a difficult art.
  - Have to discuss design rationales and rules for safe extension.
- Expensive to develop, expensive to document, expensive to learn, great saving in use.
- A functioning framework saves much because it comes with a prebuilt decision structure.
- You, too, can and will design frameworks.

Reuse

There are many excuses for not achieving reuse.
I do not accept them.

Series on Object-Oriented Design

Reuse is cost effective, but not free.

1 tested class costs 2 work weeks = $ 4,000 to develop.
A 3-class framework costs $ 12,000 - $20,000 to develop.
A purchased class library costs $ 500 - $2,000.
Therefore -->
    Spend weeks selecting class libraries!
    (it will cost weeks selecting and evaluating)

Design in Object Technology: Class of 1994

Excuse 1: Reuse is <u>very</u> hard to manage!

->How you measure productivity if the goal is fewer lines of new code?
    Are you reading comics or the class catalog?

-> Which project pays for the extra cost of a reusable part?
    Do you make a hasty one now and fix later, or
    Delay this one to make it good?

-> Who gets the reward, the inventor or user of a class?

Excuse 2: Reusable parts are hard to find, or poorly documented, tested, and maintained!

->All true.
Corporate reusable parts are available, but typically poorly documented and tested, not maintained.
    -> but inexpensive to get, still cost saving.

Industrial reusable parts cost real money, and are showing up on the market.

Excuse 3: Design with existing parts may make the design suboptimal.

-> True.

Your design will be different depending on which parts you presuppose exist.
- Can you have an "implementation-neutral" house design? or skyscraper architecture?
- Engineering design is the business of cost tradeoffs and feature tradeoffs.

Your design and parts catalog is your vocabulary.

Allocate time and Spend time in it.

Excuse 4: People are not mentally, socially trained to spend time seeking, trusting others' code.

-> True again.

Reuse is up to you.

There are too many interesting things to build to waste time rebuilding yet another transaction log or linked list.

Learn to hate redoing existing work.

The topic of reuse is full of empty words.

Just <u>do it</u>.   Yourself.

Design in Object Technology: Class of 1994

Reusable parts are "reusable" only after several "reuses"; they are typically thin, not fat.

"Reuse is free. Everything I build is reusable!"
- -- Senior designer

->What is wrong with that sentence?

"If it has not been reused, it is not reusable."
- Project leader experience report..

A first version usually contains project-specific features that must be taken out.

To make it reusable, remove the project dependencies, making it "thinner".
- do not add "generally useful features", making it fatter.

Summary: Reuse is costly but cost effective, hard to establish, hard to manage. Allocate time for it.

1. It takes time, energy and money to find and use someone else's software.
2. It takes adjustment on the part of the developer to seek out and use someone else's work.
3. The burden of effort is on the developer, not the manager.
4. Some manager still has to cost justify spending time making a class reusable.
5. All told, there is still a factor of 5:1 or 10:1 in cost savings from reusing parts over developing parts.

Design in Object Technology: Class of 1994

Methodologies

Series
on
Object-Oriented
Design

A methodology provides a framework of communication between teams and members.

1: It fosters clear communication.
- 1a. Allow another person to understand previous work.
- 1a. States deliverables and standards for them.
- Interpersonal responsibilities. Team processes.

2: It forms a basis for education.
- 2a. Guidance on use of techniques.

3: It is larger than 1 person.
- 3a. 1 person needs a technique, 10 people need a methodology.
- 3b. The job description is part of the methodology

-Hiring a person to do a job is a fact of the methodology.

Design in Object Technology: Class of 1994

Most OO published "methodologies" provide insufficient guidance, are overly cumbersome.

Most (Booch, Rumbaugh, Shlaer&Mellor, Odell) focus on design notation: a standard/ a deliverable/ a technique

Most have insufficient tool support.
   Change the code, then go change the design sheet.

Most have a process that says, "Try. Iterate".

Role statements, technique guides, deliverables guides, standards, process guides, tool guides are missing.

Each has some attractions and some drawbacks.

Learn how to appreciate their qualities.
- BUT, Do NOT feel compelled to merge them!

Humans and Technology
Alistair Cockburn   © 1994-2021 Alistair Cockburn
section (slide) L1-14 (153)

Booch attractions: rich notation, famous author, corporate backing.

o The first "object" "methodology" published.
o Sold over 100,000 copies.
o Booch committed to absorbing all notations
   Down to C++ coding level (parallel syntax?)
   Logical, physical, structural, behavioral diagramming.
o Backed by Rational Corp, with courses and tools
   Evolving capabilities.
o 1994: Booch teaming with Rumbaugh to unify ideas.

Humans and Technology
Alistair Cockburn   © 1994-2021 Alistair Cockburn
section (slide) L1-14 (154)

Design in Object Technology: Class of 1994

Booch drawbacks: too rich a notation, hard to maintain, changing fast.

- o Day 1: subset the notation
    No project can use and remember it all.
- o After day 1: devise a process
    It contains no process guide
- o Install a change control process
    Changing code invalidates the design model.
    Rational working on reverse engineering tools.
- o Be prepared to evolve
    Booch steadily learning, absorbing, changing.

OMT attractions: rich notation, compatible with pre-OO practices, gives design guidance.

- o Structural and behavioral modeling notations.
- o ER, data-flow and FSM diagrams
    People familiar with them.
- o Specific chapters in book on finding and evaluating classes, database design, etc.
- o 1994: Rumbaugh teaming with Booch to unify ideas.

Design in Object Technology: Class of 1994

OMT drawbacks: too many notations, hard to maintain, changing and out-of-date.

- o Everything but the kitchen sink; Don't try to use it all.
- o Academic; Hard for standard practitioners.
    - And getting more academic over time.
- o Process: Different projects need different setups.
- o Inadequate tools to manage changes in code.
- o Be prepared to evolve.
    - Rumbaugh steadily evolving.
    - Change with him as he learns more and better.

Objectory attractions: use cases, metamodel, relatively complete, traceable products.

- o First OO methodologist with functional requirements and formal interaction diagrams.
- o Covers requirements gathering through test.
- o Methodology includes process, even models itself!
    - Tuneable development process
- o Traceability and code generation built into tools.
- o Jacobson has 30 years of experience in telecomm.

Design in Object Technology: Class of 1994

Objectory drawbacks: cumbersome in practice, missing design guidance.

- o Usage reports indicate very heavy process
    - Constraining, cumbersome deliverables
    - "Must be good for large projects"
- o Focus on deliverables, short on techniques
    - Recently adopted responsibility-based design
- o Short experience with application development
    - Developed from telecommunications background
- o "Use cases" overused, unclear in the industry.
    - Lovely buzzword, conflicting definitions

Ptech attractions: compatible with pre-OO practices, formal, clean & readable notation.

- o "Objectified" structured analysis notation.
    - ER diagrams with inheritance
    - Data-flow diagrams with no data stores
    - Event modeling with event subclassing
- o Sophisticated and precise semantics
    - Possible to generate code and execute directly
- o Simple enough for users to read and even write
    - Demonstrated
- o Complete notation, even object reclassification.

Ptech drawbacks: no provision for reuse, hard to translate to design.

o Techniques focus on fresh development each time.
   No discussion of "design with reuse"
   No place for "design patterns"
o Fully separate function and object models
   Hard to find where the class boundaries belong.

Coad attractions: simple, complete notation.

o Single notation encompasses object interaction diagrams and ER diagrams.
   In advance of its time. May become more popular.
o Simple process: analyze a bit, design a bit, build a bit.

Design in Object Technology: Class of 1994

Coad drawbacks: simplistic, lacks supporting tools, bad design guidance.

- o Simple drawing notation requires sophisticated tools to be effective
    - Not available yet, but check around 2002.
- o Books contain naive or questionable advice
    - E.g. employee subclass of person
- o Wouldn't it be nice if developing OO systems were so easy!!
    - Missing attention to algorithms.
- o Still missing interteam processes.

Shlaer/Mellor attractions: formal, event-driven, metamodel, compatible with pre-OO CASE tools.

- o Design ideas based on their experience in event-driven, real-time systems.
    - Handles asynchronous, real-time demands
- o Complete deliverables from analysis down to code.
- o Established market with non-OO CASE tools
    - Save money - reuse the CASE tool
- o Good discussion of object life cycles.

Design in Object Technology: Class of 1994

Shlaer/Mellor drawbacks: improper CASE tool support, hard to maintain, too much analysis.

- Inadequate tools - must reenter information by hand.
    - Time consuming and frustrating
- Change in lower-level design forces rework in higher design
    - Fixable with better tools?
- Excessive time in paper analysis and paper design
    - Coding is easy from the final design document --
        if your project lives that long

Wirfs-Brock attractions: effective, simple, good design guidance.

- Responsibilities demonstrated effective on multiple projects.
    - Consistently highly rated by busy designers
    - Valid for subsystem as well as class design
- Easy to state, relatively easy to learn.
- Book provides thorough discussion of design issues.

Design in Object Technology: Class of 1994

Wirfs-Brock drawbacks: design technique only, no use cases or interaction diagrams

- No requirements, coding or test included.
  "Designing OO Software" is an accurate title
- Needs addition of use cases for requirements, interaction diagrams for documentation.
  (She is using those now)
- Good technique, needs to be complemented with other techniques and processes.

This course attractions: effective, simple, has a metamodel, design guidance, relatively complete.

- Second generation methodology, based on project debriefings from previous methodologies.
- Responsibilities as cornerstone, from Wirfs-Brock.
- Use cases as cornerstone, from Jacobson.
- Design technique gleaned from expert designers.
  Includes "design with reuse" and "design patterns".
- Metamodel ensured completeness and tool guidance.
- Implicit roles and processes.

Design in Object Technology: Class of 1994

This course drawbacks: no deliverables description, no process guidelines, no tool support.

- o Still incomplete against the methodology framework:
    - Multiple processes possible
    - No business rule modeling
    - No document standards.
- o No tool vendors with full support of: use cases, responsibilities, class design.
    - Check again in mid 1996.
    - These three likely to become the staple of Booch and Rumbaugh and Jacobson over time.

Recommendation: pick a methodology, use it, improve it, use it some more.

A methodology is personal, fitting the organization.
Choose one. Use it.
- Choose one likely to work.
- Better too light than too heavy.
- Pay attention to interteam issues.

Decide what you do/don't like, do/don't care about.
Choose conventions that work for your organization.
- The result is Your methodology.

Use it again. Tune it again. It is Yours.

Survival Tips

Series on Object-Oriented Design

(T1) Focus on the end result: adapt to whatever is needed to deliver.

OO development is different. You will want to change many things about the way you work.
- Small, deep teams
- Increments, iterations, evolutionary prototypes.
- Matrixed ownership (class x function)
- No baseline estimation curves.

Ask not for more time, ask for more flexibility rules.
- Reduced intermediate deliverables
- Incomplete requirements
- Timeboxing

Design in Object Technology: Class of 1994

(T2) Control expectations: OO is just a packaging technology that lets us think differently.

Otherwise rational people suddenly believe they can do 10x as much in 1/10 the time.

You can do more in less
* with lot of practice
* and a good reuse library

OO does not generate any code for you... it lets you reuse code artfully.

OO does not make (business model) = (code model)
* Good subclassing is an art, born of thinking, experience, luck and revision.

Humans and Technology
Alistair Cockburn © 1994-2021 Alistair Cockburn
section (slide)
L1-15 (173)

(T3) Get useful tools: #1-versioning, #2-lan work, #3-hyperlinks, ... beware partial code generation.

Versioning: the most difficult thing to do manually.
    Do it, however is needed, for everything.

Lan working eases sharing of results.
    Check-in, check-out

Hyperlinks: still mostly a wish, but making progress
    Get from any keyword to code, model, or glossary

Partial code generation does not survive iterations.

"CASE" tools: drawing editors, but expensive.
    (and do not survive iterations.)

Humans and Technology
Alistair Cockburn © 1994-2021 Alistair Cockburn
section (slide)
L1-15 (174)

Design in Object Technology: Class of 1994

Expertise: hire some or rent some, but get some.

There are so many great pitfalls, you don't need to fall into the same old ones.
There is so much to learn.
- Try "just-in-time" education

Find experts:
- OO project manager
- OO team communicator (methodology)
- Language specialist
- Class library specialist

(T5) Use cases + algorithms + responsibilities: necessary and sufficient (but not optimal)

Designers and methodologists are converging on use cases for requirements, responsibilities for partitioning.
Opinions are split on use of class diagrams.
- * Some favor complete modeling
- * Some favor occasional use

No methodology discusses algorithms.
- Algorithms are personal, methodology is for teams.

Design in Object Technology: Class of 1994

(T6) Achieve reuse: it depends on YOU trusting your predecessors!

10-20 work-days to invent, test and document a class.
.5-1 work-days to learn to use one.
> Is your ego worth loss of x20 in productivity?

Do you really want to spend your life coding linked lists?

Reuse is hard to manage:
- Who is goofing off, who is learning the library?
- How does one brag about writing 10 loc/day?
- How does one measure productivity?
- Reward the creator or the user of a class?

Reuse is up to you. Do it.

(T7) Manage your use of C++: find the key 60% and stick to it.

Moving to C++ has nearly sunk two companies and has drowned many projects.

It is a large language with many alternatives.
> Ignore 40% and use the key 60%.
> (Object message... Object message...Object message)

Problem 1: which 60% is the correct set to use?

Problem 2: How do you get everyone to use that 60%?

Get a C++ expert (a real one) to help you select the 60% for your organization.

It is up to you to stay with that 60%.

Design in Object Technology: Class of 1994

(T8) Design with shield classes: know where you are putting flexibility.

Shields are conscious design choices.
... have a match in the domain, ideally.
... protect a set of subclasses.
... package an interface with multiple variations.
... should be both obvious, and documented, at the end.

(T9) Develop incrementally: it eases mid-course adjustments.

People learn by doing to completion.
Do one increment in 4-8 weeks.
Take time to examine & adjust after every increment:
    way of work, teams, process, estimates, education, ...
Use the validation V to your advantage.

Design in Object Technology: Class of 1994

(T10) Plan time to rework: you will rework anyway.

Iteration is harder on a manager than are increments.
  It means doing something twice.
Identify and separate sections likely to change.
Plan a 2-month period to clean up and improve key areas.
  Programs rewritten a second time are shorter, faster, <u>and</u> easier to maintain.
  Better for your experienced team to do it with fresh memories, than for your maintenance team to do it without understanding.

(T11) Train your manager: increments and the difference between activity and progress.

Managers have to fix time and budget,
  and have to plan iterations within that.
They cannot support YOU if they do not understand what you are doing.
  Are you getting great reuse or reading books?
  Writing new code or cutting and pasting?

Final test: what is wrong with this statement?
- "C++ is really productive - I can write 500 loc/day"

Design in Object Technology: Class of 1994

Four stories

Series on Object-Oriented Design

Story 1: The case of too much prototyping.

Beginning: no OO background, no OO education
  "OO is prototyping - we'll prototype until we like the product, then convert it to product code."
  --medium sized Smalltalk project

Middle:
  Audience: "This prototype is too slow!"
  Audience: "This prototype is too slow!"
  Executive: "You don't take feedback. Turn this prototype into a product now, or get cut."

End: cut.

Lessons from the case of too much prototyping: use a Requirements Model, dispose of it quickly.

"Rapid Prototypes" do not evolve into code, but expectations say they do.

Create a Requirements Model, but make sure either:
- (a) it is product quality and can turn into product,
- (b) or it vanishes as soon as possible.

Diminishing Visible Returns in back half of project is hard to manage.
- Work in increments to control Diminishing Visible Returns.

Story 2: The case of Brooklyn Union Gas

Beginning: Consultants, but not deep OO ones.
- Bet-your-company, main line project
- 150 mainframe programmers
- "We'll write our own PL/1 variant to do OO"
- Large, mainframe, non-GUI, relational database.

Middle:
- ?? <-- "What did they do?!"

End:
- Successfully delivered and working
- Produced and being maintained.

Design in Object Technology: Class of 1994

Lessons from the B.U.G. case: Plan for danger, set and follow simple standards, be conservative.

Technology was considered new, likely to change. So:
1. Reduced use of inheritance because it is hard to change on disk.
2. Iterations in key areas, e.g., run time.
3. Simple standards followed consistently

Use cases and responsibilities (or things like them) work. Graphical notation is unnecessary or worse.

OO can be done on any platform, in any language.

Story 3: the case of following a recipe.

Beginning: no OO background, self-taught
 "Just model the world and turn into objects."
 large, multinational, PC / host
Middle:
 PC people in one department, host in another
  - no cooperation
 "Naive" real-world model, failed 1st deliverable.
 Converting to objects did not help.
End: Struggling to convert and save project.

Design in Object Technology: Class of 1994

Lessons from the case of following a recipe: keep your eyes open, communicate well.

Host/WS teams is necessary but dangerous. Create owners for functions and classes.
- Matrix model of ownership

Do not trust to a recipe, have to always be thinking.
Do not "just" model the world.
Only in hindsight does the design model the world.

"OO is not programming as usual"

Story 4: the case of the agile project.

Beginning: no OO expertise
large, C++ development, 2 work groups
Got permission to use increments, even on the requirements and schedule creation.

Middle:
Crashed 1st iteration. Replaced programmers. Got 3 months education. Tied managers together.
Tardy 2nd iteration. Matrix model. Everyone codes.
Timely 3rd iteration. Tuned structure and process.

End: on time, expanded to muli-site interleaved project.

Design in Object Technology: Class of 1994

Lessons from the case of the agile project: ask for flexibility, keep your eye on the deliverables.

- Focus on the end result, and adjust as you have to to get there.
- Matrix model of functions x classes works.
- Use incremental development to provide places to replace, fix, tune your operation.
- Don't ask for more time, ask for more flexibility.

Look both ways before you cross the road, then CHARGE!!   Good luck

Design in Object Technology: Class of 1994

Selected reference slides

Series
on
Object-Oriented
Design

Humans and Technology
Alistair Cockburn   © 1994-2021 Alistair Cockburn
section (slide)
L1-17 (193)

A methodology provides a framework of communication between teams and members.

1: It fosters clear communication.
- 1a. Allow another person to understand previous work.
- 1a. States deliverables and standards for them.
- Interpersonal responsibilities. Team processes.

2: It forms a basis for education.
- 2a. Guidance on use of techniques.

3: It is larger than 1 person.
- 3a. 1 person needs a technique, 10 people need a methodology.
- 3b. The job description is part of the methodology

-Hiring a person to do a job is a fact of the methodology.

Humans and Technology
Alistair Cockburn   © 1994-2021 Alistair Cockburn
section (slide)
L1-14 (194)

Design in Object Technology: Class of 1994

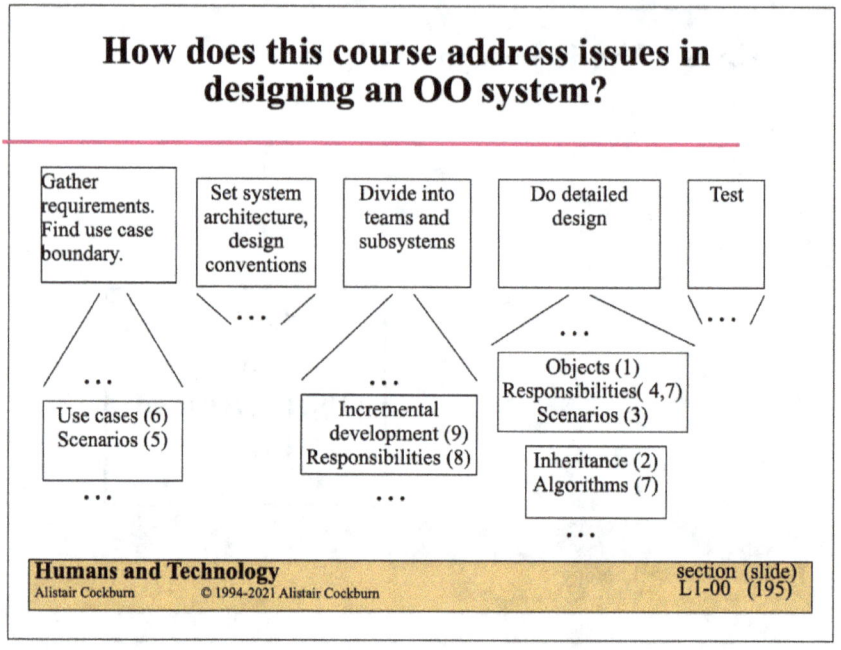

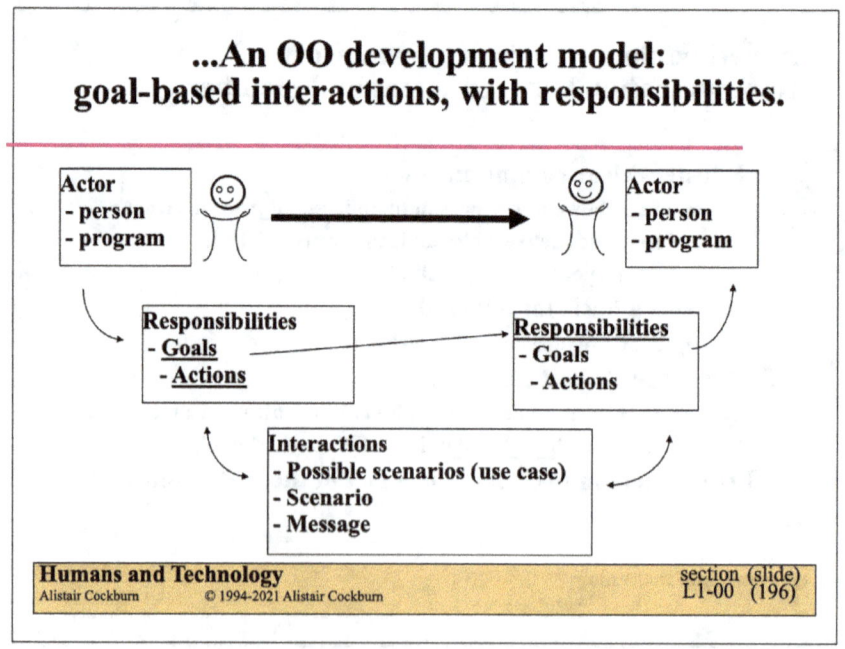

Design in Object Technology: Class of 1994

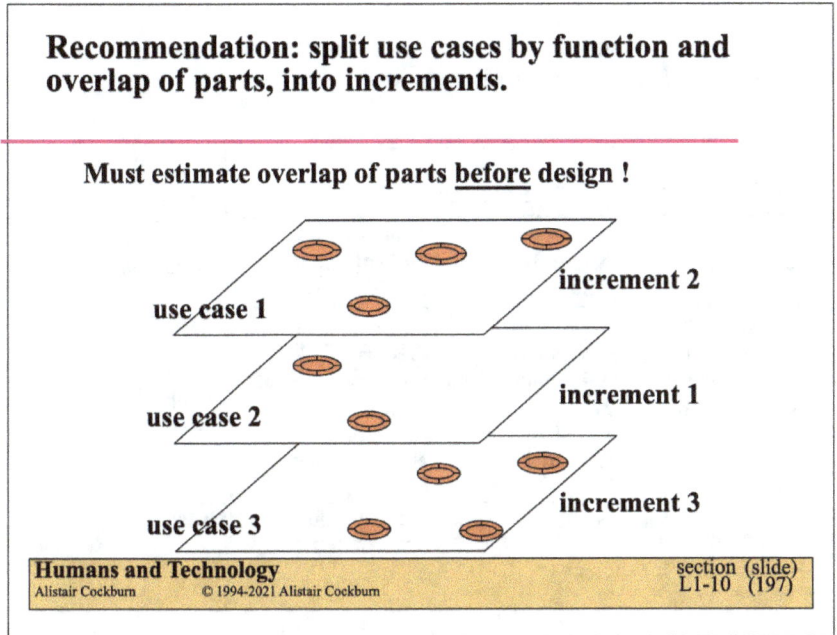

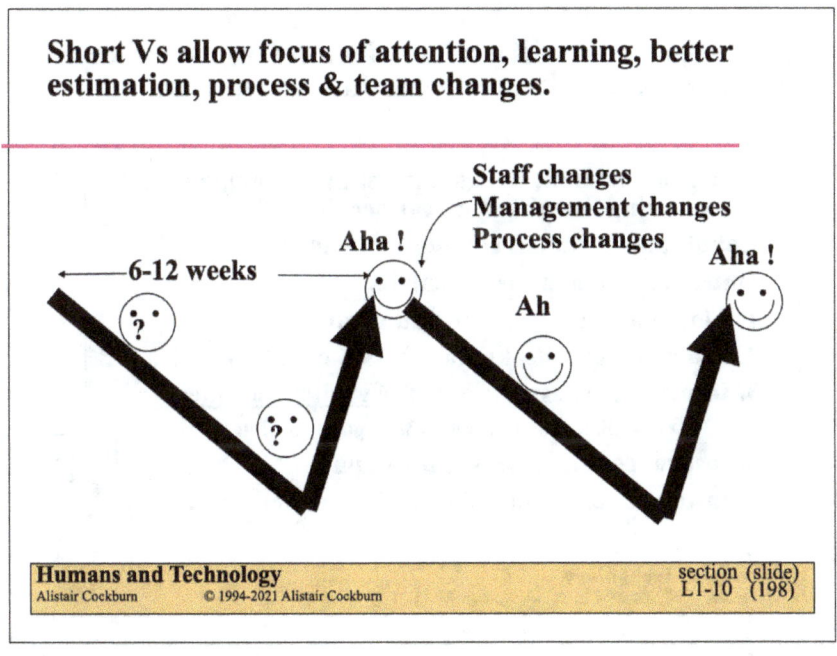

Design in Object Technology: Class of 1994

A use case has success and failure outcomes, a scenario has only one outcome.

Primary Actor: Account owner Primary Actor's Goal: withdraw money			Use case characteristic information
Knows codes, Has funds.	Knows codes, No funds here, Other funds.	Knows codes, Not sufficient funds anywhere	Scenario conditions
Presses button.	Presses button.	Presses button.	Trigger
Get codes. ok. Ask amount.ok. Give money.	Get codes. ok. Ask amount.Nok. Transfer funds ok. Give money	Get codes. ok. Ask amount. Nok. Transfer funds. Fail. Refuse money.	Scenario steps
Succeed		Failure	Outcome

Humans and Technology
Alistair Cockburn © 1994-2021 Alistair Cockburn
section (slide) L1-07 (199)

Use cases contain scenarios, which reference use cases, which use scenarios, etc. - continually smaller.

Scenario: claimant writes car insurance company requesting benefits from car accident.

Conditions: claimant eligible for benefits

Outcome: claimant paid benefits.

1. Claimant submits claim and details.
2. Insurance co. checks claimants file -- all ok.
3. Insurance co. assigns agent to <u>evaluate accident</u>
   • -- all accident details within policy outline.
4. Insurance co. <u>assesses value to pay.</u>
5. Insurance co. pays claimant determined value.

use cases

Humans and Technology
Alistair Cockburn © 1994-2021 Alistair Cockburn
section (slide) L1-07 (200)

Design in Object Technology: Class of 1994

Summarize functional requirements in 4-columns: actor, goal, system responsibility, data needs.

ATM actor	goal	system responsibility	data needs
account owner	withdraw money	give $, receipt; update balance	acct #, code, amount.
bank employee	refill cash drawer	update cash balance.	$ amount.
maintenance staff	refill paper	register paper (non)empty	paper present.
tester	test many situations	read/run test scripts, & produce report.	test scripts.
installer	initialize system	reset to start state.	"initialize" signal.

Summarize interface requirements in a table: use case, trigger, secondary actors, interface types.

use case	trigger	secondary actors	interface type
withdraw money	key press	bank central cash dispenser	database table. hot link.
refill cash drawer	lift $ sensor	cash dispenser	hot link
test many situations	program command	server computer	flat file.
initialize system	key press	none	- - -

©Alistair Cockburn, 2021   105

Design in Object Technology: Class of 1994

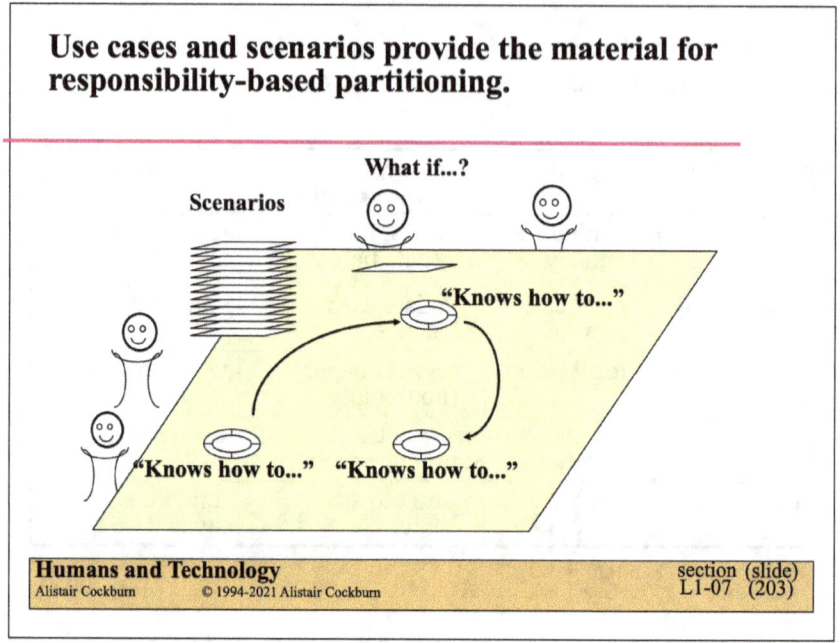

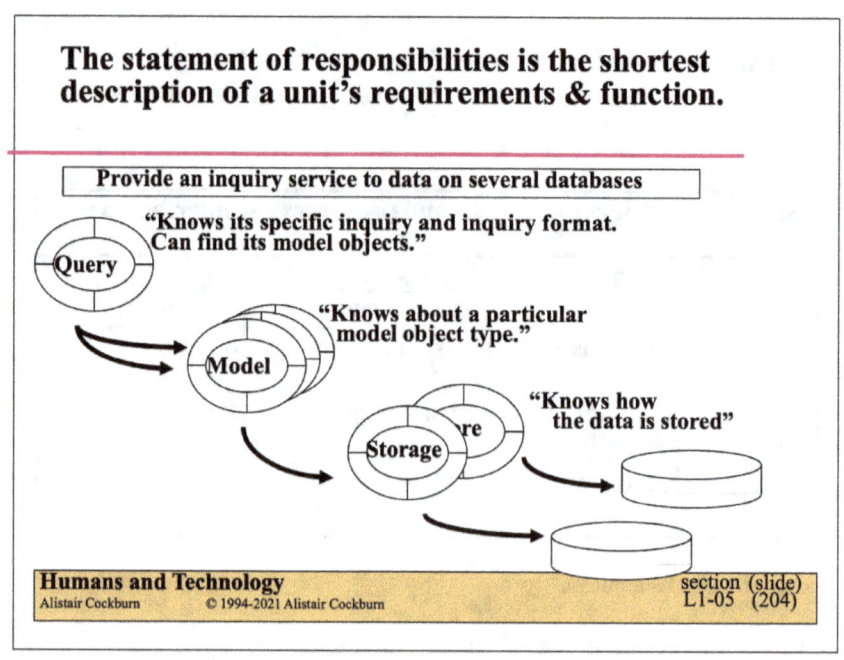

Design in Object Technology: Class of 1994

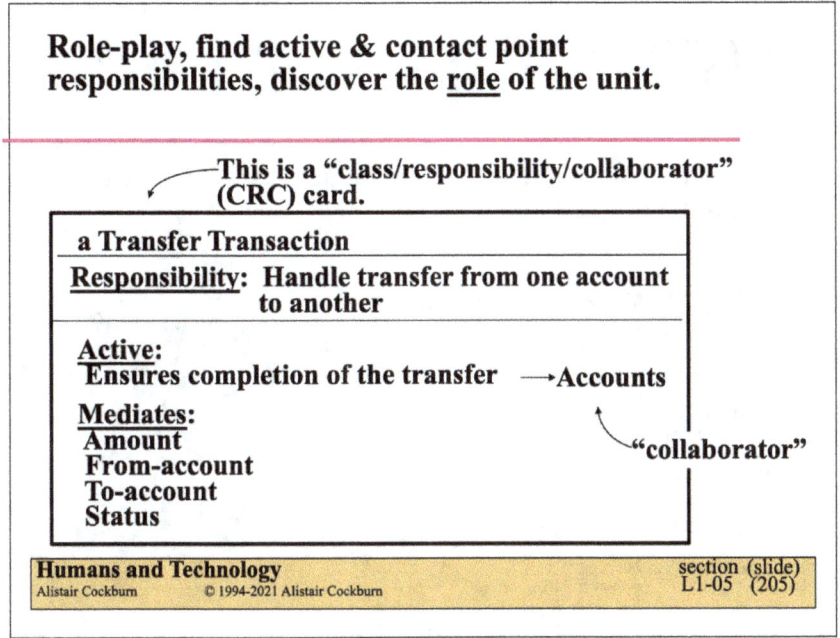

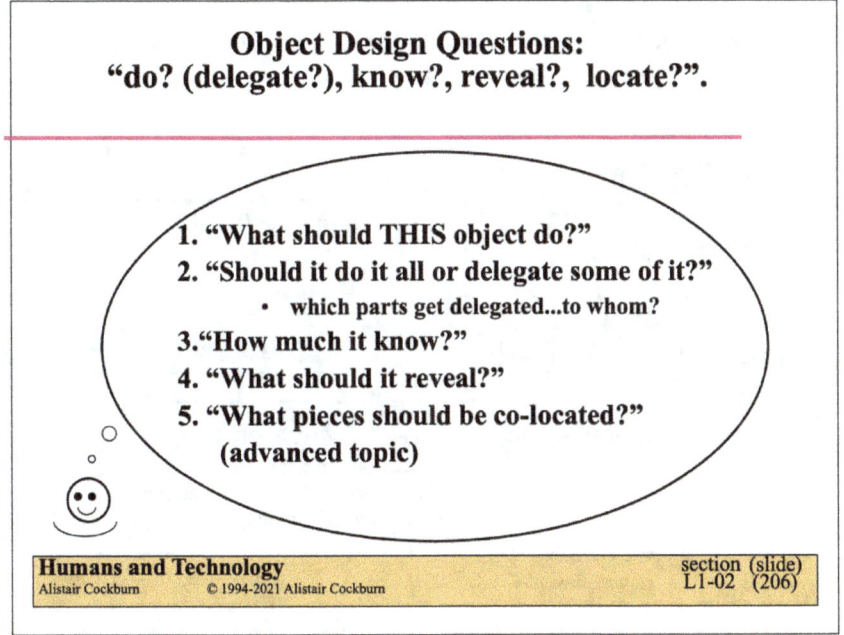

Design in Object Technology: Class of 1994

Interaction diagrams are crucial to understanding an OO design.

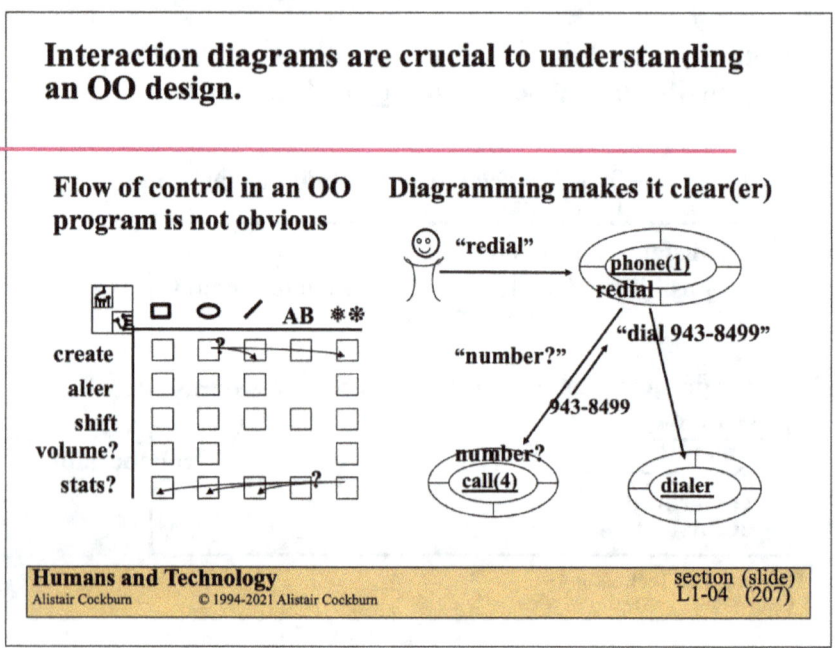

Flow of control in an OO program is not obvious

Diagramming makes it clear(er)

Typical views of an OID are top and side views; there are many ways to draw them.

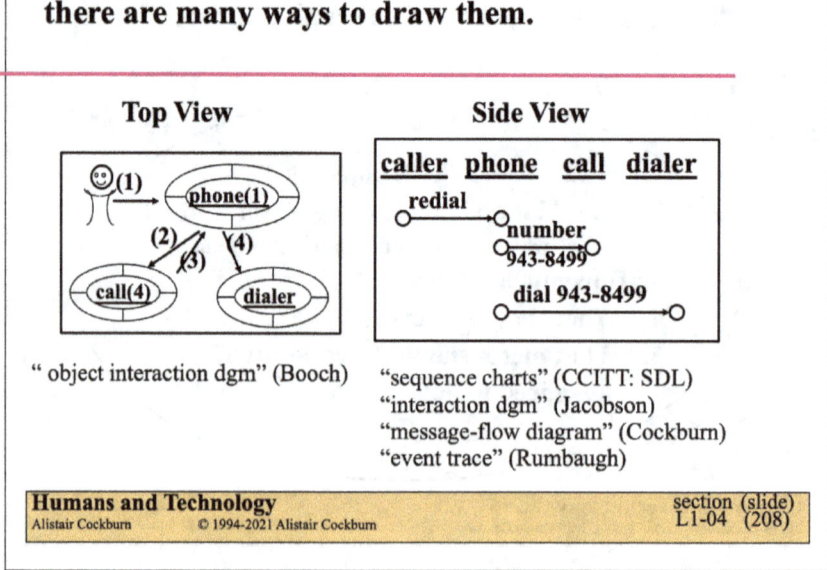

Top View

" object interaction dgm" (Booch)

Side View

"sequence charts" (CCITT: SDL)
"interaction dgm" (Jacobson)
"message-flow diagram" (Cockburn)
"event trace" (Rumbaugh)

Design in Object Technology: Class of 1994

Rule of Design: Identify classes early, but leave inheritance to language-specific class design.

1. Identify which classes are needed.
2. Identify commonality between classes.
3. Let the class designer decide how best to implement the commonality
    - Interface inheritance
    - Subclassing
    - Instance data

...A system model separating domain & application from transformers.

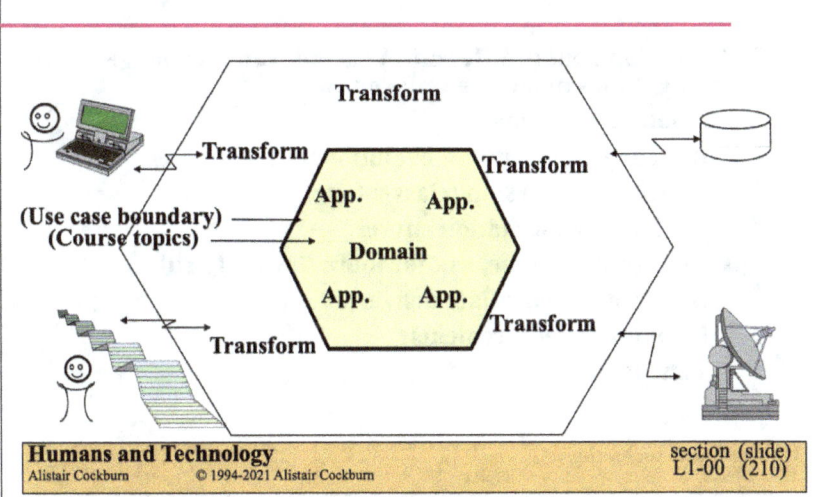

Design in Object Technology: Class of 1994

Design 2: Model and multiple interactors

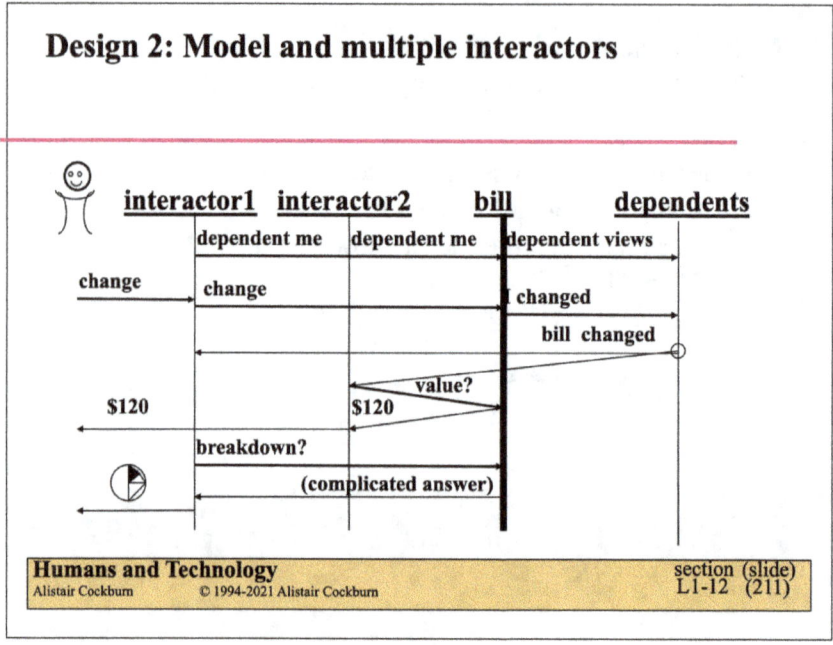

Humans and Technology
Alistair Cockburn © 1994-2021 Alistair Cockburn
section (slide)
L1-12 (211)

(T1) Focus on the end result: adapt to whatever is needed to deliver.

OO development is different. You will want to change many things about the way you work.
- Small, deep teams
- Increments, iterations, evolutionary prototypes.
- Matrixed ownership (class x function)
- No baseline estimation curves.

Ask not for more time, ask for more flexibility rules.
- Reduced intermediate deliverables
- Incomplete requirements
- Timeboxing

Humans and Technology
Alistair Cockburn © 1994-2021 Alistair Cockburn
section (slide)
L1-15 (212)

Design in Object Technology: Class of 1994

Exercises

E0. <u>Bank</u> (a "pre"-exercise)   Design in an "object" manner an entire bank branch office (not just the computer part) :ustomers, money, checks, places to put money, reports. Discuss how to document the design and document some part of it.

E1. <u>Comparing Dates</u> (messages)   Class participation: Show how two dates decide which one is greater than the other. Requires 8 people.

E2. <u>Pizza delivery</u> (objects)   Sketch a computerized system for recording pizza deliveries. Decide first on the menu for your pizza shop. Decide on the number of delivery zones to be handled. Take from order pickup to payment. Decide on the objects. Discuss how to document the design.

E3. <u>Bank accounts</u> (Inheritance)   Design one part of a banking program: different kinds of accounts. Decide what kinds of accounts your bank will support, and what variations within each. Identify the classes, possible inheritance structure, polymorphism; how to open, close, use each account.

E4a. <u>Human coffee machine</u> (responsibilities in partitioning)

Design a coffee machine out of humans, with humans carrying out all the operations. Decide what each person is to handle. Document the design with responsibility statements and interaction diagrams. First, design a simple coffee machine. When it works, add complexity and features. The instructor will add more variations. Watch how responsibilities are assigned and change with the changing assumptions and requirements. Introspect and discuss.

E4b. <u>Coffee machine requirements</u> (use cases, goals, goal failure)   Create requirements for your coffee machine: actors, goals, responsibilities, goal failures. Use the 4-column functional requirements chart.

E4c. <u>Coffee machine controller</u> (Models and shields) Design a controller for your coffee machine. Do robustness analysis. Document the design using responsibility summaries and interaction diagrams.

E5a. <u>Pricing requirements</u> (Recursive Design and Incremental Development)

Write requirements for a simply price management system. Note that the price of your goods is based on component costs, plus costs that depend on other department, and that prices change over time. Identify primary, secondary actors, the service of your system. Identify the first increment of work.

E5. <u>Pricing design</u>   Design the first increment of the pricing system. Find the key objects and responsibilities. Assume the presence of standard system services. Document the components, responsibilities needed. Check for rpbustness. Time permitting, move on to the second increment.

E6. <u>Sorted collections</u>

Design a transaction log, ordering by date, initially.   Assume a component is available called a "sorted collection", which automagically keeps things sorted according to whatever function you name. Name the function it will need. Once you have an idea how to make the log, figure out how the sorted collection works. Make it work with different sorting functions, sorting any kind of object. Class participation: test your design using people.

End: Design in Object Technology

Series on Object-Oriented Design

The end

Design in Object Technology
"Class of 1994"

Series on Object-Oriented Design

Alistair Cockburn

©Alistair Cockburn, 2021

www.ingramcontent.com/pod-product-compliance
Lightning Source LLC
Chambersburg PA
CBHW070930080526
44589CB00013B/1457